SLATE, SAIL AND STEAM

A HISTORY OF THE INDUSTRIES OF PORTHMADOG

JOHN IDRIS JONES

AMBERLEY

Dedicated to those individuals who worked so hard and skilfully in these industries, and especially in memory of those whose loyalty and commitment to their work took them to their deaths.

First published 2016

Amberley Publishing
The Hill, Stroud,
Gloucestershire, GL5 4EP

www.amberley-books.com

ISBN 978 1 4456 5347 1 (print)
ISBN 978 1 4456 5348 8 (ebook)

British Library Cataloguing in Publication Data.
A catalogue record for this book is available from the British Library.

Typeset in 11pt on 14pt Celeste.
Typesetting by Amberley Publishing.
Printed in the UK.

Contents

Acknowledgements

I would like to thank the following for their help in the creation of this book:

Denise Idris Jones
James A. Jones
Tricia Newey
Janet Guegan
Chris Preece
Derry Bryant
Simon Kenyon
John M. Wills
Stan Morton
Annwen Jones
Alun John Richards
Andrew Oughton

Introduction

This book is about people taking control of their own destiny.

These people invested their determination, skills and money into businesses which sometimes succeeded and sometimes failed.

In this book we are concerned with three industries: slate extraction and export; sailing ship manufacture, and locomotive and rolling stock engineering.

Individuals who start businesses are set on improving their lives financially and socially. When these businesses expand, they require workers, and their wages flow through their families to the advantage of their community and society.

This process was particularly striking in the Porthmadog area in the early part of the nineteenth century. In 1790 it was an area of poor land, with no settlements and virtually no employment. By the 1830s it was completely changed. Large-scale slate quarrying had started 12 miles away in Blaenau Ffestiniog. Multiple shipbuilding, driven by the need to transport the slate by sea, was growing on the seashores in and around Porthmadog. Transport from Blaenau Ffestiniog down to the port at Porthmadog was provided by the Festiniog Railway Company, which in the 1860s was the first narrow-gauge railway in the world to use steam locomotive power for commercial reasons on a long and tortuous upland route.

Starting all this activity were a number of driven and determined individuals – the pioneers.

The first was Methusalem Jones, who in 1765 assembled a gang of men in Cilgwyn quarry, Nantlle Valley, and walked to Gelli farm, in a remote part of Ffestiniog. He is said to have had a dream featuring a pickaxe being driven into land here and discovering the high-grade 'old vein'. He was the progenitor of the great Blaenau Ffestiniog industry. He formed the Diffwys quarry with a partnership of workers and so created the world's first slate-quarrying company.

The next two are William Alexander Madocks and John Williams, pioneers of the Porthmadog region. They were different in character and personality. Madocks, the son of a wealthy lawyer, was confident and extravert. Williams, the son of a poor farmer from Anglesey, was modest, patient and meticulous. Between them, in the first quarter of the nineteenth century they built two embankments, created a village and a town, rescued over 1,000 acres from the sea, turning it in to fertile land, and created in Porthmadog a harbour which received slate from up in the hills, sending it in home-built merchant ships to the markets of the world.

After that came the slate Czars: Samuel Holland Junior, Henry Archer, Lord Newborough, William G. Oakeley, Edwin Shelton, William Turner, William Casson and John Whitehead Greaves. These were English-speaking people and originated outside the area.

Slate is at the centre of our story.

Blaenau Ffestiniog was a thriving slate centre through the nineteenth century, with thousands of men employed in quarries and mines. These men were determined to improve their lot, creating a regular income for their families, working in the most dangerous of conditions, often dying early from accidents at work or lung disease.

Slate needed to find its way to market, especially overseas, so ships were built in Porthmadog to take this valuable commodity sometimes very long distances. Shipbuilding started on the seashores in and around the new town of Porthmadog in the 1820s and by the time it finished, in 1913, over 250 vessels had been built. It was an extraordinary story of self-help and mutual cooperation: shares for each vessel were purchased by ordinary people of the neighbourhood – shopkeepers, teachers, quarrymen. Ships were created by designers who were largely self-taught, their timbers sourced from local woods: local carpenters, blacksmiths, engineers, sail-makers and painters worked together, making a ship on any available piece of seashore. They were mostly schooners, very shallow in the water, about 100 feet long, slim, and unstable when not ballasted by slate, which is heavy. They had two or three masts, very tall, and huge sails. They were fast and beautiful: they were among the finest vessels ever built on British shores.

The entrepreneurs of shipbuilding who stand out are: Henry Jones, Ebenezer Roberts, Simon Jones, William Griffith, Richard Jones, David Jones and David Williams. These were Welsh-speaking people and originated within the area.

Slate needed to find its way from the quarry in Blaenau to the port, so the Festiniog railway was built to transport it. As railway enthusiasts know, the Festiniog railway is a unique survivor. These days tourists have pleasure in travelling narrow gauge up gradients, through tunnels and a spectacular landscape, from Porthmadog to Blaenau Ffestiniog. The story of the formation and running of this railway through the nineteenth century is one of triumph over extreme odds. Such a railway had never been built before. It served as a conduit for the export of slate and without it, the huge expansion of the slate industry in this part of North Wales in the second half of the nineteenth century would never have happened.

The prime movers of the Festiniog railway were James Spooner, Charles Easton Spooner, Henry Archer, Samuel Holland and Charles Menzies Holland. They were English speaking and came from outside the area. Perhaps we can help to correct the bias of history and add two more pioneers who came from the Welsh-speaking sector: James Pritchard and William Williams.

The following pages are an attempt to bring together the essential details of a unique place, period and process of industrial history. The story of Porthmadog, its industries and its rapid expansion is not replicated in any other part of the UK. The parts of it are connected. It is an account of the discovery and quarrying of slate, of carrying it down to the seashore by steam train on a narrow-gauge railway and of shipping it from there in locally built merchant vessels.

This account describes towns rising from nothing, of industries created by the ingenuity, persistence and sweat of men. It is a story of human interconnectedness and cooperation, of invention created by necessity, of businesses, industries and communities born out of a determination to advance, prosper and succeed.

1

Slate

Slate is unusual stuff. It is a natural substance, formed in the earth's crust. It is very old, hard, heavy, fine-grained, stratified and impervious. It was formed when high levels of clay and silt sediments were impacted by volcanic activity: it is a metamorphic rock. It was formed, essentially, out of fire, heat and then intense pressure. It retains the essential form of the sedimentary, being stratified, and thus being able to be cut in straight lines.

It was used by the Romans in the third century as flooring when they built their fort – Segontium – at Caernarfon. In the thirteenth century it was used by Edward I in the construction of his castles. Bishop William Morgan (translator of the Bible into Welsh) in the late sixteenth century specified slate for the reroofing of St Asaph cathedral. In 1932 when this roof was being repaired, the 400-year-old slates were examined, found to be in good condition, and used in the reroofing.

Slate has always been valued for its water-repelling quality. It now sits on roofs in buildings across the UK, in Europe, especially Germany, as well as in North and South America – in Buenos Aires, Argentina, and, it is said, on the roof of the courthouse in Boston, Massachusetts. It will stay in situ virtually forever; nobody knows how long a slate will last. Getting it from its source to faraway places is a subject of this study.

In origin, we are talking of the period represented by three geological periods – Silurian, Ordovician and Cambrian. The first two are the oldest periods of the Paleozoic Era. Silurian marks the period 438–408 million years ago. Ordovician marks the following period until 443 million years ago. After that is the Cambrian period – its oldest part is indicated at 590 million years ago. All three periods are represented in the slate deposits of North Wales. The Cambrian deposits run across north-west Wales with three quarrying areas located in an arc some 6 miles from the coast. The Penrhyn quarry of Bethesda is inland from Bangor, the Dinorwic quarry of Llanberis inland from Caernarfon and the Nantlle Valley quarries, notably the Dorothea, 8 miles south-west of Llanberis. (The Cilgwyn quarry in the Nantlle Valley dates from the twelfth century and is one of the oldest in Wales.)

The Ordovician period is named after the Celtic tribe, the Ordovices. It was named by Charles Lapworth in 1879 because Adam Sedgewick and Roderick Murchison were placing rocks from the Cambrian and Silurian period into the same period. He based his views on observation of fossil fauna. The Cambrian period was first identified and named by Adam Sedgewick (1831–5) who studied the rock formations of North Wales. These rocks are mainly sedimentary – sandstone, shale, limestone – formed in shallow seas. Slate stands out because it is the product of volcanic activity. The Cambrian rocks are notable as the first to contain easily recognised fossils, especially the Trilobite. The name Cambrian is recognized by geologists across the world, and it is interesting that it was first discovered and named from the study of the old rocks of North Wales.

The Ordovician deposits run from south of Betws-y-Coed to the shoreline of Beddgelert/ Porthmadog. It is these that form the huge deposits in the Moelwyn and Migneint mountains which cradle Blaenau Ffestiniog. Slates from this area make the best roofing slate. Another band of slate from the Ordovician period runs through Corris where there was significant quarrying.

The earlier Silurian period is represented by quarries around Machynlleth, including Abergynolwyn, inland from Tywyn on the west coast.

So in the west of North Wales, there were, and to an extent are, six principal slate-quarrying locations. However, there were hundreds of smaller quarries. A tour through the Horseshoe Pass, out of Llangollen, will reveal slate waste coming down to the edge of the main road, and a number of slate quarries, one of which is still productive.

Through the Middle Ages, and in newer centuries, slate was quarried in small-scale quarries. In 1413, the area of the later Penrhyn Quarry was recorded in a rent-roll of Gwilym ap Gruffydd with a number of his tenants being paid 10 pence each for working 5,000 slates. In the fifteenth century Guto Glyn asks the Bishop of Bangor to send him a shipload of slates to roof a house at Henllan near Denbigh. Slates were exported to Ireland in the sixteenth century; the wreck of a sailing boat from this period carrying them was unearthed from the Menai Strait. There is a record of production from the Penrhyn estate (Bangor) from 1713 when fourteen shipments totalling 415,000 slates were sent to Dublin. Looking at the elegant roofs of Dublin today, you can observe Welsh slates.

Before the slate boom in the nineteenth century, slate in North Wales was quarried by local men who would pay a royalty against sales to their landlord. They would use horse and cart to carry it to a nearby port and ship it in small boats to England. After the turn of the century, landlords began to run slate extraction themselves. Under single ownership, the Penrhyn and Dinorwig quarries became the largest slate quarries in the world, and in Blaenau Ffestiniog, under the ownership of the Oakeley family, the Oakeley mine became the largest slate mine in the world. In 1898 in north-west Wales, some 17,000 men produced half a million tons of slate.

In the middle of the nineteenth century, the value created by slate extraction came to about half the gross domestic product of the whole of north-west Wales, the area now known as Gwynedd.

Men who worked in the quarries (*chwarel*) were Welsh speaking. When they looked at vertical rock in the quarries, they called it *y wyneb* (the face) or sometimes *y garreg* (the rock). A finished slate was called *y llechen*; many slates were *llechi*.

The decline of slate production in north-west Wales coincided with the famous labour dispute in the Penrhyn quarry between 1900 and 1903. Subsequently, the decline continued when the First World War took thousands of men to battle.

Roofing slate continued to be in demand but in the 1960s and '70s it was in competition with cheaper roofing material from abroad, particularly roofing slates and tiles from Spain, so the period saw the closure of smaller quarries. Today, slate is still quarried in North Wales on a small scale and new uses for the remarkable substance are being found, such as for worktops in kitchens and grate surrounds.

There is no doubt that working in a slate quarry, or underground in a mine, was an exhausting and hazardous occupation. Exploitive, iniquitous practices prevailed. Men worked for six days. Their wages were poor. They would not become a full miner until they served five years as an apprentice. Even when fully qualified, they were not paid a regular wage but had to endure the indignity of 'piece-rates': these were based on monthly contracts, with the first three weeks paid on a lesser wage, with the remainder being made up in the fourth week of work unit profits and bonuses. However, a much-resented burden to the workers was that they were forced to pay the cost of their working tools and essential items themselves, such as ropes, tools, sharpening and air for pneumatic drills. Rock faces were cut into by men who wrapped ropes around their bodies, dangling down the vertical rock, using as the tools of their trade, hammers and chisels. As the slate and detritus was often wet, ropes sometimes broke or became slippery and men fell to their deaths.

The first process in slate mining was to free a piece of rock from the face. This might weigh hundreds of tons and was done by means of explosives. The quarrymen would drill into the rock in specific places and fill the hole with dynamite or 'black powder' (*powdwr du*). After wiring the explosives together, men would withdraw to a safe distance and detonate the explosives. This was a procedure filled with hazard and many men were blown up by the explosive powder they had laid, or by unexpected rock fall.

It is no accident that on the shores of the river Dwyryd, adjacent to Minffordd, Penrhyndeudraeth, just over the mountain from Blaenau Ffestiniog, was a factory of *Cook's Explosives*, owned then by the Nobel company. And on a spit of land between two beaches between Borth-y-Gest and Criccieth, in a suitably remote location, there is a stone building originally built for explosives storage.

Quarrymen and quarry miners worked on a 'bargain' system. 'A bargain' was a piece of rock about 6 metres square. Men were paid according to the quality of this unit, and bargaining took place between the workers' representative and the steward employed by the owners. A usable block of slate, removed from the mother rock, would be about 100–200 kg each. This block would be cut into rectangular slabs which were then cut to different thicknesses. The final process was splitting the slate to a few millimetres in thickness, a very skilled operation using a sharp wide-blade chisel and a mallet of African oak.

There are many sizes of slates. The largest is the 'Empress' at 26 by 16 inches, and the smallest, 'Narrow Ladies', at 14 by 7 inches. Between these sizes, all with female, and colourful, names are ten other sizes. A standard roofing slate would be the 'Dutchess' at 24 by 12 inches. However, a roof incline would be made up of different sizes, the larger ones across the base and the smaller ones across the top, presenting an attractive gradation; this being the traditional and best method of slate roofing.

Illness in quarrymen was widespread. Lung diseases were common, as in the coal mines of South Wales where silicosis infected the lungs of thousands of the workers in damp, airless and dust-strewn atmospheres. Medical help was scarce and expensive and miners did not have the money to pay for it. The UK's National Health Service ('free at the point of use') was born out of these conditions, in the coal valleys of South Wales, championed by a local Member of Parliament, Aneurin Bevan.

There is no doubt that slate mining has left a negative legacy on the natural environment of Wales, particularly north-west Wales. The problem is waste. Slate mining is very man-intensive and product-intensive. For every usable unit of slate blasted from the hillside, then sawed, cut and cut again, there is a quantity of at least ten units of waste created. Traditionally, what was there to do with it? Dump it. Around every slate quarry there are mounds and mountains of slate waste. The road called 'Crimea', now the A470 which runs the length of Wales from Llandudno to Cardiff, runs down to Blaenau Ffestiniog between huge masses of slate waste. This material is gradually being taken away and used as footings for civil engineering projects and buildings, as hard content in the making of artificial stone, as decorative stone for garden design and even for cosmetic products, and so on. It will take a long time before the unsightly waste from slate quarrying and mining will be cleared and the landscape returned to its natural state.

2

Porthmadog

Madocks/Williams

The town was new and was named Portmadoc in the 1820s. It was a time of English influence and the 'c' reflects the name of its founder, William Alexander Madocks. Today we use the Welsh form, with the 'g' reflecting the softer, easier to say aspect of Welsh spelling and pronunciation.

Madocks, however, was a Welshman. His ancestors occupied the Dyserth-Tremeirchion area to the east of Denbigh in the Vale of Clwyd. They go back 600 years, and the name Madoc appears often in the family tree. The first to use Madocks was John Madocks of Bodfari (1601–62): his youngest son, also John, was sheriff and capital Burgess of Denbigh. This John married his second wife, Jane Williams, heiress of Fron Yw, a house on the eastern edge of the Vale of Clwyd with a glorious view over Denbigh and half the vale. This is where William grew up.

William was born in 1773 and his father was a barrister. He had a busy practice in Loncoln's Inn and the Middle Temple, London, and became 'one of the most eminent King's Counsel in England', according to Elizabeth Beazley in her splendid biography *Madocks and the Wonder of Wales*. William was his third son; his brothers were John Edward and Joseph. He was born in London but spent most of ten years at Fron Yw and then he was sent to Charterhouse boarding school. From there, at the age of sixteen he moved to Oxford as a student at Christ Church, becoming a fellow of All Souls and for eighteen years he was Member of Parliament for Boston, Lincolnshire. He was a friend of Tom Sheridan, the playwright's son. His family had friends and relations in Wales and William often visited them. One notable connection was with Sir Watkin Williams Wynn of Wynnstay, Ruabon, known as the King of Wales. Sir Watkin owned thousands of acres of good land, so his income from tenancies was considerable. William's interest in country estates was stimulated and was extended when his brother John bought a plot of land at Erith, at the mouth of the Thames, and built himself a house on it. Below the house, on the river's edge, was a piece of land which grew corn; this land had been drained in Elizabethan times.

His father died in 1794, and in his will he said that £50,000 Consolidated Bank annuities were to be held by trustees, one of which was William. The instruction asserted that the money was to be used for the purchase of freehold land.

He bought Dolmelynllyn, Ganllwyd, north of Dolgellau. He and his friends enjoyed the oak-dense valley and the Rhaeadr Ddu (black falls).

The draining of the inlet of the sea beyond the mouth of the River Glaslyn had been mooted by Sir John Wynn of Gwydir (between Llanrwst and betws-y-Coed) in 1625. He wrote to his friend 'The honored Sir Hugh Myddleton, Kny, Bart' in London describing the project and asking for his help. The request was politely declined. Hugh Myddleton said that he had experienced public works in 'seeking of coals for the town of Denbigh, within less than a myle from where I had my first being' and that he was 'full of business ... the river at London ... my weekly charge being above £200, which maketh me very unwilling to undertake any other worke. [which would] require a man with a large purse.' He never said a truer word!

The critical year was 1798. Madocks came with friends to Maentwrog and turned towards the sea. They were in the Aberglaslyn Pass. The word means seaside place of the blue lake. The sea came up to near the bridge over the river. Ahead of them was Traeth Mawr, the 'great sands', which was an inlet of the sea and impassable except at very low tide and very dangerous for travellers making their way from the south or east to the Lleyn Peninsula or Caernarfon.

This, Madocks must have thought, is a great place for landscape improvement. If he could drain Traeth Mawr he would make a fortune, gaining thousands of acres. He knew that by law, if one owned land adjoining the sea, the land beneath the water, when drained, would become his property. He was in luck. Eight holdings on the Caernarfonshire side of the estuary were to come on the market, including Ynys Madog, where the ancient Prince Madog is alleged to have sailed from and discovered America: the similarity in names was a pleasing coincidence. The main part of the land for sale centred around what was later called Portmadoc, including Ynys y Towyn (island on the seashore), located across the road from the later Festiniog railway station, and Ynys Cyngar (island of Cyngar where, according to *Gestiana*, page 33, Cyngar ab Geraint, grandson of Constantine, Duke of Cornwall, landed about the year AD 600), now housing the caravan site Garreg Wen (white rock), on the Morfa Bychan (small seashore) road:

> Ynys Cyngar was originally surrounded by water and before the construction of the Cob and harbour at Traeth Mawr was an anchorage for ships. It was from here towards the end of the last century that shipments of slate from the Brynyfoel Quarry were made to Ireland and other places. Vast areas of Morfa Bychan on which houses are now built were once under the sea. (*Gestiana* p. 25)

The other parcel of land was situated where Tremadog now stands, stretching almost to Penmorfa (end of the seashore), including the elevated farmhouse Tan-yr-Allt, which Madocks altered and occupied. It was occupied by Shelley, the poet, in 1812/13 and later by the Greaves family. Below the Madocks residence there was a smaller house called Tan-yr-Allt Isaf, which was later demolished.

The main road from Lleyn to England passed through Penmorfa, then out of the tide's way to Beddgelert. Another smaller road or track went over Traeth Mawr to Merionethshire and South Wales: this was dangerous and frequently caused loss of life.

William bought the estate and immediately planned an embankment which would claim 1,000 acres of land; this was started on 22 March 1800. This first enterprise was 2 miles long and stretched from near where Ysgol Eifionnydd is now to beyond Prenteg, near the mouth of the Glaslyn. The Rector of Penmorfa wrote,

> This meritorious undertaking has been a great blessing to the country, particularly during the scarcity of 1800, when above 200 poor men were in constant employ. And kept off the neighbouring parishes, greatly to their relief and comfort, while all around were starving.

This first embankment was from 11 to 20 feet in height. It was made of sand covered by 11 acres of turf. Madocks wrote: 'I am now at Penmorva (sic) amidst 150 wheelbarrows and 200 spades and hearts too all attempting Canute like to set boundaries to the Ocean.' The land reclaimed was 1,082 acres and in 1801 oats were growing there. The Board of Agriculture awarded the project its gold medal.

This is where we introduce John Williams. He was born on 9 May 1778, the son of William Williams, farmer, of Ty'nllan in Llanfihangel Ysgeifiog, Anglesey. John was first employed as a gardener at Plas Newydd, home of the Marquis, and when he came to Tremadog looking for work, he was first given the job of tree planting and garden creation. His assiduous attention to detail and ability to work hard attracted Madocks's attention and soon he was made agent over all the works. He was Welsh-speaking and Madocks realised the importance of this if he was to acquire and persuade local men to work for him to the advantage of their community. It was through John Williams's hands that all wages were paid. He married Anne, sister of the late David Williams of Castell Deudraeth, and they lived in Tyhwnt I'r Bwlch, overlooking Porthmadog, next to Madocks's house Morfa Lodge (which he never occupied). He was an ardent member of Tremadog church. He died in 1859 at the age of seventy-two and was buried in a vault below the church. Here also are buried his wife and his only son, W. T. Massey Williams. An inscription in the church reads 'endowed with a strong mind and equally strong affections, he secured the good opinion of the public and the deep regard of his friends.'

If Madocks conceived of Tremadog, Porthmadog and its harbour, and the draining of Traeth Mawr, John Williams was instrumental in making it all happen. Madocks was away in London and Lincolnshire much of the time, as an MP, but John Williams, his faithful agent, was on the ground locally, organising.

Madocks designed Tremadog, giving its two roads the names London Street and Dublin Street. He had intended this place to be the stopping point for traffic going from London, via Shrewsbury, down the Pass of Aberglaslyn, to Porthdinllaen on the Lleyn Peninsular, which he intended as the embarkation point for Ireland. This part of the plan was never realised, the route via Telford's bridge and across Anglesey, departing from Holyhead, being preferred. By 1808, due to the persistence of John Williams in securing builders and tenants, the central part of Tremadog was completed, including the town hall with a theatre on its upper floor. On 3 August 1808, Sheridan's *The Rivals* was performed there; ticket costs were 3s 6d for boxes and 1s 6d for the gallery.

Writing in 1856, a local diarist (probably Owen Morris) writing under the name Madog ap Owain Gwynedd (the prince of Gwynedd who was alleged to have discovered North America in the twelfth century) penned the following in an essay for a local society:

But perhaps the most wonderful of Mr Madocks' doings was the erection of a Factory, capable of containing 60 looms for broad cloths, where all the improved machineries in that branch of manufactory were used, together with a Dye-house and Fulling-mill. We have no means of learning the amount of business done at the Factory, but it must have been considerable while it lasted. By the end of 1809 there were 68 houses and 303 inhabitants in and about Tremadoc.

This gifted author also tells us, surprisingly, that in 1799 at high tide the land which became the footing for Tremadog was 9 feet underwater!

In 1806, the idea of the second embankment began to occupy Madocks's mind. The plan was to build a 21-foot-high embankment, 1,600 yards long, enclosing 3,042 acres, so that the

Caernarfonshire and Meirionnydd shores would be joined. It was originally envisaged as a route of communication, but during the building of this embankment it evolved into something else. It would enable a railway to be built across it, bringing slate from the Minffordd shore, which had come down the Dwyryd river, to Porthmadog instead of the cumbersome present process of taking slate downriver in barges and loading it on to seagoing vessels off Ynys Cyngar in the Glaslyn/ Dwyryd estuary. This second embankment took from 300 to 400 men to build it.

In this early period, the idea of a railway stretching up the 700 feet to Blaenau Ffestiniog, 12 miles away, was not taken seriously. It was thought impossible.

It was not until the 1830s, after the narrow-gauge rail was laid down between the two towns by the Festiniog Railway Company (note the single 'f'), that slate was carried on rail all the way. It used the power of horses to pull the trucks uphill, with the gradient bringing them down. Horse power associated with steam engines did not come on until the 1860s.

In 1807 Madocks steered an Act through Parliament. It allowed him to build the embankment and to keep the drained land of the estuary to himself. He bought the farm Penrhyn Isa (the lower peninsula) on the Merioneth side in order to quarry its rock. Work started in March 1808.

By the end of 1808 work was well underway; quarries at both ends supplied stone which was tipped into the sea. From 300 to 400 men were constantly employed, and in 1810 there were 104 horses at work. Madocks thought the work would take under two years at a cost of £23,000. A tramway was built at both ends and stone was carried on wagons drawn by horses. By the quarry on the southern shore, Penrhyn Cottage was built, incorporating workshops, offices and stabling for the horses – it was occupied in January 1809. In May 1811 it was renamed Boston Lodge in honour of William Madocks and his career in politics.

Hundreds of loads of rock a day were tipped into the water. The more this happened the faster the tide ripped through the remaining gap. Acccording to Owen Morris, 'to overcome the difficulty of closing the gap, several vessels loaded with stone were built, and an immense quantity of piles fixed.' It was frustrating work. At the completion, the embankment was 90 feet wide at its

The second embankment. This can be seen to the left of this picture. The old county of Meirionnydd is in the background, with the old country of Caernarfon in the foreground. A quarry face can be seen behind Boston Lodge, the railway terminus. This estuary is here the mouth of the River Glaslyn.

underwater base; 21 feet clear of the water and 18 feet wide at the top. It was an outstanding example of landscape engineering.

In July 1811, after three and a half years of toil and with creditors circling, the 1,600 yard-long embankment was completed. In September 1811 an Embankment Jubilee was held in celebration, with an ox roasting and an *eisteddfod* (a competitive poetry and singing festival). Horse racing was held on Morfa Bychan sands. The *North Wales Gazette* reported it with verve: the event included 'our Cambrian Nobility, Clergy and gentry'. The Madock Arms tavern was busy and the town hall had a celebratory ball. The *Gazette* said that meals provided by Madocks were 'a most sumptuous collation ... of the choicest description'. An 'ordinary' (a meal supplied for the general public) held in the Madock Arms was 'crowded beyond all description' and was overseen by Madocks himself. Samuel Rogers, the diarist, was in Tremadog for the Jubilee; he wrote that Madocks was, 'a great Lord in his little city of Tre Madoc'. However, not all were happy. The novelist Thomas Love Peacok, living in Maentwrog, disliked the scheme. In 1816 he published *Headlong Hall*, with its Squire Henry Headlong closely resembling William Madocks. He wrote:

> A scene which no other in this country can parallel, and which the admirers of the magnificence of nature will ever remember with regret, whatever consolation may be derived from the probable utility of the works which have excluded the waters from their ancient receptacle ... The mountain-frame remains unchanged, unchangeable: but the liquid mirror it enclosed is gone.

This second embankment (soon to be called the Cob) was estimated to cost £23,500. It actually cost in the region of £60,000, and William Madocks was deep in debt as a result.

On 14 February 1812 the winds and the tides turned against the Cob and tore a large hole in its centre. John Williams set about organising gangs of men to effect repairs. Letters were sent asking for men, carts and horses and the replies from local farmers were positive. Self-help was

The embankment (Cob) is shown on the Porthmadog side with a train about to enter the station. The houses in the foreground are built on the old shipbuilding site, Rotten Tare.

evident. Sir Thomas Mostyn sent 151 men and 150 horses; Lord Bulkeley supplied fifty of his men. Shelley (1792–1822), then renting Tan-yr-Allt, came out as supporter and fundraiser. He was an early married man in 1813 and the locals took against him and his young wife Harriet. Here he continued work on his long poem *Queen Mab*. He and John Williams attended a meeting of the Corporation of Beaumaris, where Shelley made an impassioned speech for 'this great, this glorious cause.' The subscription list came to £1,185, of which the poet promised £100. An annual rent of £100 was set for the Shelleys. However, he wrote some unflattering comments: 'Welsh society is very stupid ... They are all aristocrats or saints ... [there is] more philosophy in one square inch of any tradesman's counter than in the whole of Cambria.' It is not surprising that he had no friends among the Welsh. It was said that a break-in to the Shelleys' residence was organized by local shepherds who objected to the poet's alleged habit of using his pistol to kill seriously ill sheep he encountered during country walks.

Harriet's initial enthusiasm for the Madocks project cooled, writing, 'We are now living in his house where formerly nothing but folly and extravagance reigned ... Here they held midnight revels insulting the Spirit of nature's sublime scenery.' On 3 March 1830 Shellley wrote to his publisher that he was the subject of an 'assassination': the story was that at night an intruder had forced his way into Tan-yr-Allt and taken a number of shots at the poet. Harriet said that part of his nightgown was singed. They left for Dublin shortly afterwards.

In 1814 the repair work was completed. Madocks was seriously in debt. In 1818 he married a widow from Breconshire and his life and finances improved.

On 9 December 1814 Madocks wrote to John Williams:

I assure you I employ my mind incessantly in thinking how to compass those important objects necessary to complete the system of improvements in Snowdonia, any one of which wanting, the rest loose half their value. If I can only give them *birth*, *shape*, and *substance* before I die, they will work their own way with posterity.

In another letter, he writes:

Thank God! The embankment is safe – that is the *keystone* to everything, to our success, our triumph, our security, our glory, and our emancipation from difficulties.

A visionary, maybe, but certainly a master with the English language.

Another piece of luck presented itself. The Glaslyn, having been diverted, was rushing through the sluices at Ynys Towyn, carving out a deep hole between the island and the shore. This gave Madocks his second big idea. He would create a harbour here big enough to take merchant ships so that slate could be carried away from here rather than from the ships anchored off Ynys Cyngar. So what started as an idea of communication, which was behind the building of the Cob, continued to an idea of trade and international expansion. So the harbour at Porthmadog was born.

In 1821 one slate quarry dominated the trade. It was worked by Messrs Turner & Casson, shipping slates to the value of about 10,000 tons per annum.

Slate was carried in trucks over the Cob, drawn by horses. Samuel Holland founded his slate depot inland from Rotten Tare in 1824 (where the Festiniog railway station and the row of small houses facing the harbour are now). Tolls were placed on goods brought across the Cob and on 'loading from the Quay of Ynys Towyn – that was then the name for the new town. In 1824 the harbour was ready to receive vessels of 60 tons. Slates brought down the Dwyryd came across to the new harbour at Porthmadog instead of unloading into ocean-going vessels off Ynys Cyngar.

Griffith Griffith and his four sons took on the huge task of sourcing stone, dressing and laying it, and creating sound walls to surround the harbour and create quays. John Williams was officially styled, 'Director of Works and the Harbour of Port Madoc', so in 1825 the new name for the place had emerged and trade began to flow.

The deeper water of the harbour encouraged shipbuilding. One of the first vessels to be built in Porthmadog (1824) was the *Two Brothers*, a smack (single-masted, fore-and-aft rigged) owned by three shareholders described as mariner, slate-loader and farmer, built by Henry Jones on a beach known as Y Tywyn, surrounded by gorse. The two brothers were William and Daniel Parry. This site, on the north edge of the new harbour, where the Maritime Museum is now, later to be Oakeley Wharf, was known as *Canol y Clwt* (centre of a piece of land).

Henry Jones was the main early shipbuilder; his snow (later brig) *Lord Palmerston* (111 tonnage, square stern, carvel planking; two equal masts, five square sails on each) was built here in 1828. Her design was based on that of the *Gomer*, built at Traeth Bach in 1821. She originally was financed by her sixty-four shares being held by fourteen men, ranging from farmers, merchants and gentleman, including Griffith Humphreys of the City of London. (A snow is a small ship carrying square-rigged main and foremast with a smaller mast behind the main mast carrying a fore-and-aft sail set with a gaff and boom.)

Between 1824 and 1830 eleven sailing ships were built in Porthmadog.

Sad to say, on an extensive European tour with his wife and family, William Alexander Madocks died on 15 September 1828 in Paris. He was buried in Pere Lachaise cemetery, in a plot now overgrown and unmarked. The people of Porthmadog initially did not believe it, so soundly had his name and personality been stamped upon the neighbourhood. The house he had planned, Morfa Lodge, above the port on the Morfa Bychan road, he never enjoyed. His trusted friend, John Williams, had a house built from a farmhouse close to it, Ty Hwnt I'r Bwlch, the one echoing the

Porthmadog harbour as it is today. The background behind the bridge is a marshy area which before Madocks was under water. The River Glaslyn enters the bay at the far right. The Festiniog railway station is in the centre of the picture.

other as the two were in real life. He supervised the activities of the harbour, looking down on the new town, its life and activities, which he had a major hand in creating, until his death in the early 1850s.

The population of Ynyscynhaiarn, which included Porthmadog, Tremadog, Borth-y-Gest and Morfa Bychan, which amounted to only a few cottagers and farmers in 1795, increased to 2,347 in 1851. It was 3,059 in 1861, 4,367 in 1871 and 5,506 in 1881 (its highest point).

The present bridge ('Bont Newydd') which carries the main road through Porthmadog was constructed in 1851, as was the new sluice gate at the head of Llyn Bach or Inner Harbour. The plan to turn this section into a second working harbour never materialised.

Between 1848 and 1878, 146 seagoing ships were built at Porthmadog and Borth-y-Gest.

In 1836 the tonnage of slate carried across the Cob to the harbour at Port was 4,275. In 1882 it was 120,426.

These statistics speak of the success of the work of two remarkable men.

3

Porthmadog Town, Location and Geography

Porthmadog today is a town with a population of 4,185, according to the 2011 census. It can be described as 'The gateway to the Snowdonia National Park'. However, because of provisions allowing for local business and industry, when the park was created in 1951 the larger area of Porthmadog and the whole of the mining acreage of Blaenau Ffestiniog were omitted.

It is located at the 'armpit' of the Lleyn Peninsula, at the join where the east–west southern coastline of Lleyn links with the north–south axis of the western shore of the main body of North Wales. It sits at the estuary of the rivers Glaslyn and Dwyryd, the Glaslyn coming down from the north via Beddgelert and the Dwyryd coming in from the east via Ffestiniog. The inlet fronting Porthmadog has traditionally been called Traeth Mawr (large beach). The mouth of the Dwyryd, passing the Italianate village of Portmeirion, is Traeth Bach (small beach). Between the mouths of these two rivers is the headland and town of Penrhyndeudraeth, which in Welsh means 'headland with two beaches'.

It is located in the county of Gwynedd, which has its administrative centre in Caernarfon. An older name for the district in which it is located is Eifionydd. It contains the terminus of the Festiniog railway, which runs across its high street, travelling up to Caernarfon, in addition to the original route through Penrhyndeudraeth to Blaenau Ffestiniog. It encloses a busy harbour, a home for yachting.

Its population expands from March to September with an influx of tourists, day trippers and overnighters, most of whom stay in caravans. Morfa Bychan, close by, contains a number of caravan sites, with hundreds of caravans behind Black Rock Sands, a 2-mile stretch of beach between Borth-y-Gest and Criccieth.

The geographical area presented here stretches eastwards about 12 miles from Porthmadog to Llan Ffestiniog and Blaenau Ffestiniog. Across the Glaslyn there is a fine view of Cnicht (2,265 feet), sometimes dubbed 'the Welsh Matterhorn'. The two mountains Moelwyn Mawr and Moelwyn Bach (the former 2,527 feet) stand between Cnicht and Blaenau Ffestiniog. The mountains around Blaenau form the watershed between the River Lledr flowing to the north (a tributary of the River Conwy) and the River Dwyryd flowing to the west. Following the River Dwyryd out of the estuary by Porthmadog, we come to the Vale of Ffestiniog and the village of Maentwrog, where we observe the fine stone façade of the Oakeley Arms. To the right are the mountains Garnedd Iago, Arenig Fach (2,259 feet) and the Migneint. Blaenau Ffestiniog lies below Moel Penymaen, with the mountain Siabod (2,861 feet) behind it. A couple of centuries ago, the area was virtually inaccessible, apart from on foot or horseback.

In the distance can be seen the mouth of the River Dwyryd. At this point it joins the Glaslyn. The trees to the right sit on an island which was created in the nineteenth century from ballast dumped by returning ships.

The middle years of the nineteenth century saw the population of Porthmadog double. It made a name for itself in 1851 by hosting the National Eisteddfod (a singing and poetry festival). It was held in a pavilion on ground which was underwater before Madocks came. Local poet Ioan Madog (blacksmith and inventor) won a prize with his verse 'Cowydd to the memory of Robert ap Gwilym Ddu'. *Gestiana* (p. 176) reports on the establishment of banks in Porthmadog. This is important because it tells us that money was in circulation, due to prosperity, indicating the success of local industries. Two banks were established in Porthmadog in 1837. The North & South Wales was started in High Street, in a house close to The Ship & Castle, then the National Provincial in the house of the late Hugh Hughes, the baker. Tremadog had seen the first Post Office in the area in 1805, and in 1832 one was set up in Porthmadog.

The development of the town of Porthmadog between 1795 and 1856 was so quick that many of the facilities present in the average town were missing, or slow to be installed. Sewers and drainage were a problem, planning of streets and new houses was rudimentary, and schools and medical facilities had to be created by private subscription. Local office worker Owen Morris wrote of poor public facilities;

Unpaved, undrained, and, until very recently, unlighted streets, and also the neglect shown of forming sanitary regulations for keeping the town in a decent and decorous aspect, and for the defence of Public Health.

Morris also recognized the problem of a town which had risen out of a recent surge of industrial activity, as this created an excess in the artisan class. He wrote that the new town was without

> the infinity, diversity and distinction of classes so characteristic of the society of an old established town ... The texture of the population of this place is very different – it has had not sufficient time to settle in to such nice distinctions, and peculiarities. All of its classes being engaged in trade, and mixing in mutual intercourse – the boundary lines separating them are not easily traceable or definable, and hence they have an appearance of cohesiveness and uniformity not often observable in older communites.

This astute observation by Morris was part of an essay, *Portmadoc and its Resources*, intended for an Eisteddfod competition in Salem Chapel, published in 1856. It was a time when the present Welsh-only rule for written submissions to Eisteddfodau did not apply.

4
Blaenau Ffestiniog: Birth and Growth

Historically, like Porthmadog, Blaenau Ffestiniog (the name means 'the cliffs above Ffestiniog') developed with the slate industry. In the eighteenth century, all that was here was mountain terrain and a few small quarries.

Today, its population, including a wide hinterland, is 4,875, according to the 2011 census. At the height of the slate boom in the later nineteenth century, the town's population had risen to 12,000. Michael Senior's account of the town, in his book *Portrait of North Wales*, is not flattering;

> The Crimea Pass tilts down ... steeply, into what might easily be mistaken for the entrance to the Underworld. The earth has been turned inside out, the normally hidden inside of it lying everywhere in massive heaps. It is too impressive to be ugly. It is monstrous, horrible, almost unbelievable; but so black and big as to be in some way fascinating. Out of this turmoil the town of Blaenau Ffestiniog was born. (*Portrait of North Wales*, p. 102)

This, however, is one view. Another view appreciates the drama of the landscape, its close evidence of persistence and determination among the quarrymen and miners, its sturdy stone houses in neat terraces offering reliable living spaces, its sense of community, its history of Welsh Nonconformity in religion, and the Welsh language spoken in the streets. Over 50 per cent of its inhabitants speak their native language, compared with 19 per cent for the whole of Wales. The Estyn inspection of Ysgol y Moelwyn in 2014 noted that 82 per cent of its intake came from Welsh-speaking homes.

Slates had been produced from the Blaenau area of Ffestiniog as long ago as 1575. These came from small quarries worked by a handful of men and they supplied slate for roofing local cottages and farm buildings.

The beginning of larger-scale quarrying in Blaenau (as it is called locally) came in the 1760s when men from the Cilgwyn quarry near Nantlle started quarrying in Ceunant y Diphwys, which is north-east of the present town in a hollow between the Manod and Moelwyn mountains, on the Manod slope. Quarrying was at the Diphwys site and was begun by Methusalem Jones and William Morris, two quarrymen from the Cilgwyn quarry, in the Nantlle Valley, some 6 miles south of Caernarfon. In 1769 several other quarrymen joined them, also from Cilgwyn. This is on the site of the later Diphwys Casson Quarry. Eight men, under the leadership of Methusalem Jones (who said that he had dreamed of a walk across the mountains to this site), took a lease on Gelli Farm and started quarrying. They struck the Old Vein, which produced slates of the finest quality in Blaenau, which was close to the surface at this point. They started selling slate to a wider area, using carts for transport. Diphwys's lease was sold in 1800 to William Turner (who had previously worked a small quarry in Ireland) and William Casson, quarry managers from the Lake District, who expanded the workings. For the early years of the century, this was the only medium-scale quarry in the area. It was busy and prosperous until 1830, by which time other quarries had emerged in competition. By 1856 the Diphwys quarry had been in scale operation for fifty-five years and was yielding 2,000 tons of slate annually.

Close to the Diphwys, the Cloddfa Lord quarry, started in 1791 and named after the owners of the property, the Lords Newborough, continued production on a minor scale, until the first Lord Newborough put more money into it in 1801. This lord died in 1807 when the quarry working here lapsed until 1823 when it was restarted by the second Baron Lord Newborough, son of the original owner. He spent around £15,000 through two years of working the quarry, until in 1828 he leased it to Messrs Roberts of Caernarfon, who pushed on more vigorously. After working it for five years and spending £30,000, they failed in 1833, the quarry then reverting to Lord Newborough. In 1834 it was let on lease to Messrs Shelton & Greaves, and soon after it was successful.

Another early-years quarry, Manod, began production in 1802–4. It experienced a number of owners until James Meyrick bought the lease in 1845. He developed the Craig Ddu Quarry alongside and output increased substantially.

William Oakeley was a key owner of land in Blaenau. He owned the Tan-y-Bwlch estate from 1789 to 1811, then his son William Griffith Oakeley took over, until 1835. The Oakeleys lived (for only part of the year) in the splendid mansion, TanyBwlch, over the River Dwyryd, near Maentwrog, now owned and occupied by the Snowdonia National Park Authority. In 1819 quarrying began at the Oakeley-owned Rhiwbryfdir Farm (this Welsh word means 'hill of insect land'), Allt Fawr, Blaenau Ffestiniog. This was in the hands of Samuel Holland Senior, a Liverpool-based owner of other mining and quarrying interests in North Wales, including copper. His son, another Samuel Holland, sold these workings to the Welsh Slate Company and in 1827 he opened another quarry higher up (Holland's or Oakeley Upper). In 1838 the Rhiwbryfdir Slate Company created the section known as Oakeley Middle. These three mines were busy and profitable for decades and became the first customers of the Festiniog Railway Company. By 1840 surface quarrying was becoming exhausted and underground mining started. The Oakeley Quarry was created in 1878 by W. E. Oakeley and it became the largest underground slate quarry (correctly a mine) in the world; it had twenty-six floors with a vertical height of 150 feet.

Close to the above, the Cwmorthin Quarry was started in 1826/7 by Thomas Casson and his partners, but it produced little slate until the 1880s.

These quarries were located in a curve around the head of the Vale of Ffestiniog; inside this curve the new village of Blaenau Ffestiniog was born, some 3 miles from the existing Llan Ffestiniog.

The quarries Diffwys, Lord, Manod, Holland, the Welsh Slate Company and Rhiwbach used the River Dwyryd to carry their slates to market from 1800 to 1868; they used seven quays along the river bank near Maentwrog. Their maximum quantity shipped was in the late 1830s, but the quantity shrank significantly in the early 1840s, showing the effect of the Festiniog railway.

The River Dwyryd is only some 6 miles long. It begins at the confluence of two mountain rivers and descends rapidly to the Vale of Ffestiniog, which is only some 60 feet above sea level. The river then meanders for 4 miles below Maentwrog bridge, which was the early tidal limit, as far as Briwet Bridge (Minffordd): Maentwrog originally had its own quay and small boats would load here. Nearing Pont Briwat, the river expands and becomes Traeth Bach (little beach), the cove of Abergafran (where ships were built) is on the right, part of Minffordd, and the Italianate village of Portmeirion. The Porthmadog Bar marks the line of the sea at Tremadog Bay and then Cardigan Bay.

The first half of the nineteenth century saw Blaenau Ffestiniog expand as quarrying and mining needed more and more workers. Quarries at Llechwedd, Maenofferen, Votty and Bowydd opened and became productive; the Turner and Casson quarry becoming large-scale. Dozens of other quarries in the wider areas started and flourished.

Liverpool slate dealer Samuel Holland put his son, another Samuel, in charge of his slate holdings in 1821. Selling one quarry and opening another, his output in 1831 was 650 tons a year. Holland's slates were the first to be carried by the new Festiniog railway.

In 1825, companies intent on exploiting the Welsh slate market were formed in London. One of them was styled The Welsh Slate Company. It was headed by Lord Palmerston and Lord William Powlett. They bought land and workings from Samuel Holland. In the mid-century this company employed 450 men, with an output of 19,000 tons a year.

In 1928 Samuel Holland's new quarry on the Rhiwbryfdir site grew substantially and in 1856 employed over 300 men, with a yield of 10,000 tons a year.

The local diarist Owen Morris (writing under the name Madog ap Owain Gwynedd) writes,

> The three principal works form one great cavity. The scene is so interesting and novel that we are tempted to describe it. Along the divisions, or galleries, in to which the hollow is divided, are seen trains of wagons moving backwards and forwards. On the sides are seen men suspended by frail ropes, over awful precipices, busy boring holes for blasting. Other men are engaged loading rubbish and slate slabs into the trucks, which are lifted up from the lower parts by means of water balances, and then drawn by horses through dark levels, emerging at length at the rubbish heaps. But what means the sound of the horn? At once there is a general escape, as if from the scene of some disease. Then follow the blasts – pop – pop – pop – until the very ground rocks – the surrounding mountains long reverberating the reports. Immense masses of rock are loosened – the slate rock rather gently, but the granite and bastard rocks with greater force. The bugle is again sounded, the men reappear from behind projections, strange holes and clefts, and with alacrity resume their work. The clinking of hammers is once more heard throughout the extent of the great cavity. (*Portmadoc and its Resources 1856*, p. 44)

All the early quarries were working one vein, which follows the contour of the ground, from Diphwys and Lord to Llechwedd (Mr Greaves) then to the three – the Welsh Slate Company, Rhiwbryfdir Company and Samuel Holland's quarry.

So by the mid-century 1,500 men were employed by the slate industry in Blaenau Ffestiniog. Wages to the total value of £70,000 a year were paid out.

There was a problem of getting the slate to market. Set in a cradle of mountains, 700 feet above sea level, originally road connections were poor. Quarry-owners started building houses for their workers near their place of work; one settlement was at Rhiwbryfdir, serving the Oakeley and Llechwedd quarries. Population expanded rapidly and by 1851, 3,460 people lived in the new town of Blaenau Ffestiniog.

Slate export originally was via the River Dwyryd. Slate was carried on their backs by mule or pony down the mountainside to Congl-y-Wal (the corner of the wall) at the edge of the Vale of Ffestiniog. Here they were transported by horse and cart to jetties on the River Dwyryd (some can still be seen). They were loaded onto barges and taken down river, joining with seagoing vessels in the lee of Ynys Cyngar (where the Garreg Wen caravan park, close to Morfa Bychan, is now.) This hill was originally surrounded on three sides by the sea, with a river on the Porthmadog side making it an island. Ships would draw in off its Criccieth side, where the Porthmadog Golf Club is now. The work of transporting slate down the Dwyryd and raising it onto seagoing vessels was done by a group with the name 'Philistines'. These outlandish people are reported to have their own clothing style, including tall hats. Loading slate from a barge in the deep water of the estuary, off Ynys Cyngar, was tricky, exhausting and life threatening. This export route could not compete with the directness and utility of the new Festiniog railway, even if it was horse-drawn and had gravity on its side. The horses were brought down with the train in a specially designed and built horse-dandy, attached to the rear of the train, which was an open high-sided vehicle with four sides. The railway was 13 miles and sixty-two chains long, with a difference in feet above sea level of 700. It was confined to carrying goods, but with the advent of steam locomotion in 1863, a huge improvement in the transport of slate came about. The Dwyryd route continued to be used after the opening of the horse-drawn Festiniog railway was initiated in the 1830s, but on a reduced scale, the slate being carried by the Philistines to the new Cob from the river by horse and cart.

The story of John Whitehead Greaves is worth telling. He was the fourth son of a Warwickshire banker and was on his way to emigrate to Canada. He had walked from Liverpool to Porthmadog when he heard of the prospects in slate mining. He leased a small quarry called Chwarel Lord in Blaenau Ffestiniog. In 1845 Greaves and Shelton gave up on the quarry Bowydd: Greaves took a lease on land owned by Lord Newborough but was not successful, although he was helped financially by his banker brother. He is said to have had a dream that a rich vein of slate lay on a hillside known as Llechwedd. He took years to buy out all the leaseholders and in 1846 his men started prospecting. He had married, with young children, and was now financially poor. Several places on the hillside were dug out but no quality slate was found. He was about to give up when his foreman and two other workers agreed to carry on working for a month with no pay. One evening in 1846 a horseman was heard trotting up the drive at Tan-yr-Allt. Their father told his children that the news would bring them ruin or fortune. The house door was opened and the foreman was there, on his knees; 'Now is the time to praise God,' he said, 'We have struck the old vein.' A pickaxe had penetrated the Merioneth Blue Vein, the richest there was. They had discovered such a cache of slate that Llechwedd mine was opened up immediately and out of it came some of the world's best slate, roofing houses and public buildings all over the world. In 1862 John Greaves won a medal with a piece of slate 10 feet long and 1 foot wide – it was an extraordinary $\frac{1}{16}$-inch thick and it would bend without splitting.

The 1860s and 1870s were boom times for slate production across north-west Wales. Blaenau Ffestiniog had its first school and chapel. In 1881 its population stood at 11,274. However, the 1890s saw a decline and some quarries went out of business. The two world wars took men away from quarrying and in the first half of the twentieth century mines amalgamated and closed. Cheap Spanish slate was having an impact. In 1946 the Festiniog railway closed. The once huge Oakeley Quarry closed in 1970, and Maeofferen finally closed in 1978. Llechwedd continued to work slate but its tourism side, with two runs for visitors, one to the caverns below ground, is thriving. Seeing how a mountain was mined from the inside, by means of huge caverns, one below another, is a unique experience. The revived Festiniog railway is a busy tourist attraction, its twisting, scenic route up to Blaenau Ffestiniog attracting thousands of visitors a year.

Now, in the twenty-first century, Porthmadog is thriving because of visitor numbers, and Blaenau Ffestiniog, in a sense, is reborn. Tourism has become the new industry. The narrow-gauge railway with its charming old trains has become a major North Wales tourist attraction, now extended across the arm of the Lleyn Peninsula all the way to Caernarfon, a scenic and dramatic trip indeed.

5

Oakeleys

The name Oakeley is the oldest name associated with ownership of land in Blaenau Ffestiniog and the area's slate industry.

For two centuries the Oakeley family occupied Tan-y-Bwlch, now called Plas Tan-y-Bwlch. However, it was not their main home. This was Cliff House in Leicestershire, and they also had a house in London. The name Tan-y-Bwlch in Welsh means 'Below the Gap' or 'Opening'. Plas means mansion. There are references in the early years of the nineteenth century to 'Tan-y-Bwlch Hotel': this refers to the stone building situated on the main road out of Maentwrog on the Porthmadog road. It is now called the Oakeley Arms Hotel – it was called this when the estate was sold in 1915. This part of the Vale of Ffestiniog is called Tan-y-Bwlch.

It is indeed a mansion now: tall, stone built, set high over the Vale of Ffestiniog, half a mile on the Porthmadog side of Maentwrog, with a spectacular view over the River Dwyryd and the northernmost mountains of mid-west Wales. It is now owned by the Snowdonia National Park Authority and is used as an archive and study centre.

The Oakeley's association with the house (then much smaller) goes back to 1789 when Margaret Griffith, heiress of the Tan-y-Bwlch estate, married William Oakeley, a wealthy man from Staffordshire. The estate had expanded over the previous two centuries and acquired land in the Maentwrog and Ffestiniog area. The will of Robert Evans in 1602 referred to Tan-y-Bwlch, so there was a building there then; his son's wife inherited Rhiwbryfdir, an area of land in Blaenau Ffestiniog, which proved slate-rich in the nineteenth century when full-scale quarrying started, including the huge Oakeley Quarry.

According to the then laws of inheritance, Margaret's assets became the property of her husband, so William Oakeley came into the possession of the Tan-y-Bwlch estate and ran it from 1789 until his death in 1811. He was known locally as Oakeley Fawr (Big Oakeley) and regarded with favour as he seemed fond of the area and had the well-being of his tenants and workers in mind. He improved the land below his house, adjoining the Dwyryd, which was a tidal estuary, by building an embankment which contained the sea. He started mining for slate on the family land and when his son took over he expanded their slate interests significantly.

William's son William Griffith Oakeley, who took over the estate in 1811, was a pivotal figure in the growth of slate extraction in Blaenau Ffestiniog. He expanded his father's quarry, improved roads, built quays on the Dwyryd so that slate could be better loaded on to barges, and had a hand in the formation and development of the Festiniog railway.

Part of Plas Tany-y-Bwlch, one of the houses of the Oakeley family. It is now a study centre for the Snowdonia National Park Authority.

In 1818, Samuel Holland Senior took a lease on Rhiwbryfdir farm, owned by W. G. Oakeley, for a yearly rent of £150 and a royalty of one tenth of the value of all slates raised and manufactured. The royalty amounted to between £6,000 and £7,000 per year in the mid-century.

William Griffith died childless in 1835, and the estate was then managed by his wife Louisa Jane. On her death it passed to William Edward Oakeley, a cousin's son. Louisa Jane was an active landowner and industrialist initially, encouraging the growth of Blaenau Ffestiniog and improving conditions for its residents, including building a hospital. However, her health and mental faculties deteriorated and William Edward Oakeley, looking to inherit the estate, tried to get her declared insane. In 1868 she left the area, signing over the running of the estate to William, although she remained the owner until her death in 1879.

William Edward Oakeley took a great interest in the details of the estate and built the distinctive stone buildings which fringe the main road. He extended his home, putting in the bay windows at the front. He built most of Maentwrog in a distinctive style, including remodelling the church and building a new school. He continued land improvement work along the Dwyryd; the Royal Society of Arts in 1897 awarded him a medal for his land improvement work. He died in 1912.

However, his running of the family slate business was not so successful and financial problems ensued. His son Edward de Clifford enjoyed an eighteenth birthday celebration when his father paid for hundreds of guests to be transported by the Festiniog railway. Later, he paid for his workers and their families to visit Llandudno by train.

The 1890s saw Tan-y-Bwlch supplied by electricity from its own hydroelectric source. A Pelton wheel was involved; the powerhouse was situated on the hill behind the house.

William Edward's two children inherited the estate. Edward de Cliford inherited the mansion and part of the land; his sister Mary inherited the other part. Edward spent most of his fortune in gaming and partying in London. In 1915 he sold the estate for £25,000. He died a bachelor in 1919.

The Oakeley line came to an end when Mary died at ninety-six in 1961.

Samuel Holland, in his *Memoirs*, refers to 'The Tan-y-Bwlch hotel' in the 1820s and 30s. This makes sense because the hotel building by the main road was located in the area called Tan-y-Bwlch. On 12 February 1870 Charles Easton Spooner presented a paper on the narrow gauge railway at the 'Tan-y-Bwlch Hotel'. *The Book of North Wales* by Charles Frederick Cliffe (published in 1850) described Tan-y-Bwlch as 'an extensive hotel and posting house, delightfully situated a little above the village of Maentwrog on the north side of the Vale of Festiniog, below Moelwyn.' He apparently had not realised that the name belonged to the area, not just the hotel building. The name Oakeley Arms Hotel started in the nineteenth century and continues to this day.

In 1969 Merioneth County Council bought the Oakeley's house and the grounds. It is now called Plas Tan-y-Bwlch; the property came under the control of the Snowdonia National Park Authority in 1975.

The view from Plas Tan-y-Bwlch. It shows the roofs of the village of Maentwrog. The Ffestiniog Valley stretches up to the left. The River Dwyryd runs through here.

6

Samuel Holland and Henry Archer

Samuel Holland was born on 17 October 1803 at Duke Street, Liverpool, the son of Samuel Holland who had interests in the lead, copper and slate industries of north Wales. After schooling in England and Germany, he joined his father's business as an office boy. When he was eighteen his father sent him to look after a new quarry which he had opened at Rhiwbryfdir in the parish of Ffestiniog. This was in March 1821 and was the beginning of a long association with Blaenau and Llan Ffestiniog, Maentwrog and Penrhyndeudraeth. He occupied a number of houses in these places, with his longest period of residence being at Plas y Penrhyn in Penrhyndeudraeth, which he enlarged. He died in Wales during the December of 1892.

[based on *The Memoirs of Samuel Holland One of the Pioneers of the North Wales Slate Industry*, published by the Merioneth Historical Society, 1952]

When Samuel Holland first visited Blaenau Ffestiniog, he saw two or three men at work making slates for local cottages. Mr William G. Oakeley owned the land and Samuel Holland Senior applied to him for a Take Note, which was permission to work there, and this was granted for three years. He was appointed as foreman Richard Jones from Llanllyfni. Quarry work started in 1818. When Samuel (Junior) came to take over the management of the quarry in 1821 he came from Liverpool in a Steam Packet that plied between Liverpool and Bagillt, stayed the night in St Asaph, then walked to Ffestiniog via Llanrwst, staying at the Eagles Inn and Dolwyddelan. He also stayed the night in Penmachno but could not sleep because 'rats were running about the room'. In Ffestiniog he stayed at the Pengwern Arms; arriving there, the landlady Martha Owens told him that he was expected and that he should proceed to the quarry where his father would meet him. He walked towards the quarry and came across two men discharging slates from their horse and carts. They told him they were carrying slates from Mr Holland's quarry. He met up with his father who told him he was expected the previous night. The quarry was only a small hole with only a few men at work, under the foreman Richard Jones. They returned to Ffestiniog where they had a late dinner and went to bed. They expected to meet with Mr and Mrs Oakeley, who were expected there with a nephew of Mr Oakeley's (Mr W. Edward Oakeley) who was coming from Staffordshire to see Tan-y-Bwlch. Samuel Holland Senior told his son to take over the running of the quarry and to find a decent house to live in. This was 25 March 1821; the son was eighteen years old. His father left on horseback for Liverpool. Samuel found lodgings at a cottage (Pen Mount) where the landlady kept one servant and a girl, the niece of the servant. He arranged to stay there and be served breakfast, luncheon, dinner and tea and have his washing done, for a sum of 12s per week. He seldom required the four meals. He walked up to the quarry each day, sometimes with some bread and cheese in his pocket. He returned in the evening and sometimes enjoyed a meal of mutton chop.

The slates that we made at the quarry were carried in little carts to a wharf called Pen Trwyn y Garnedd on the river side – about a mile or a little more below Tan y Bwlch hotel – so I frequently walked from the quarry to the wharf and then back to Pen Mount being in all the distance from Penmount to the quarry 3 miles, from the quarry to the wharf 6 miles and from the wharf to Pen Mount 3 and a half miles, in all about 12 miles – and this I did frequently for successive days besides walking about the quarry ... Near to the wharf was an old cottage, which had once been used as a turnpike cottage, but then not used – the turnpike having been abolished – this I thought might suit me as a residence and being very near to the wharf and on the road to Port Madoc, where I had to go occasionally. I took it at low rent from Mr Oakeley to whom it belonged; made some additions and alterations to it – built a small stable (two stalls) and decided to go there to live, which I did before the end of 1822 much to the grief of Mrs Griffiths my kind old landlady – who would, I believe, have liked me to remain on, even at half the price. I was very comfortable there and called occasionally to see the old Lady who did not remain very long after I left, but went with her servant to one of the Arms Houses at Corwen where she eventually died. I used to call and see her, when I passed through Corwen. I started then at Pen Trwyn y Garnedd a housekeeper on my own and so had to look out and engage a servant, thinking that one who spoke very little English was desirable as I wished to learn Welsh. A young woman was recommended to me ... but we did not get on very well together. I gave her a letter recommending her to my mother in Liverpool; she turned out to be a good servant ... I engaged Miss Williams at nine guineas a year; she remained with me for nearly fifty years in the various houses I occupied, and never asked for an advance in wages. Three of my cousins from Knutsford visited me (Mary, Bessey and Lucy Holland) on their way to Barmouth; they made an excursion in my cart, one horse, up to the quarry. A girl from Maentwrog used to come early in the morning to help my housekeeper, Mary Wynne. My father came over once or twice to see me and my mother and her servant Mary Richardson.

Samuel's ability to walk long distances is illustrated by his account of walking from the Dwyryd to Liverpool:

Over the mountains to Dolwydelan, then to Llanrwst, from there to St. Asaph and on to Holywell and bagillt – the first day got some bread and milk for supper and then went to bed after a walk of 54 miles that day got a boat to cross the Dee early (6 o'clock) in the morning to Parkgate, walk from there to New Ferry got a boat (with Will Chatterton) to cross the Mersey to the herculean Pottery and to my father's house near there, between 8 and 9 o'clock to breakfast, the cost of my journey was 3/6 [three pounds and six pennies], the boatage over the two rivers, (I was a great and good walker at the time – I used to walk from my cottage to Caernarvon 23 miles – go about the town – & walk back in the afternoon so doing 46 miles).

I got the quarry roads improved and got some of the farmers to get wagons instead of carts to carry the slates from the quarry to the wharf (Pentrwyn y Garnedd) that was near my cottage from where they were boated to Port Madoc or to the vessels lying afloat in the river at Port Madoc (no quays were built at the port then). My father did not consider the boats then in use of a good construction so he had two built in Liverpool, to carry more tonnage and these he sent round from L'pool under the care of one old sailor W. Jones, the first one was ready then he walked back to Liverpool and brought the other. My father had a small sloop to cry about 20 tons – built in Liverpool called *The Experiment*.

As the boats employed to carry the slates down to Port Madoc, from the slate wharves, could only be employed or worked during the Spring tides, that many nights during the Spring tides I had to be up, helping load the boats & see them off for Port Madoc. My wharfman was frequently (almost daily during the Spring tides) down at Port Madoc, seeing to the boatmen properly discharging the slates from their boats into the vessels and properly storing them. I had to see the loading of the boats and that the slates were counted properly and correctly into them.

1825 was a very speculative year; many companies were formed, for working mines and quarries everywhere, not only in Great Britain, but all over the world.

In March 1825 Samuel Holland and his father sold their Rhiwbryfdir quarry and farm for £28,000; also sold was Holland's cottage, the wharf and warehouses on the Dwyryd.

This transaction shows the enormous profits that were to be gained at the time through successful slate quarrying.

Samuel Holland took a lease on Plas y Penrhyn, Penrhyndeudraeth, and lived there for many years. In 1828 he began quarrying on the ground adjacent to his old quarry, which he had sold to the Welsh Slate Company. Mr W. G. Oakely extended his lease. In May 1827 Holland drove a tunnel in what was disputed ground, and litigation followed. Holland won his legal case and he continued quarrying, which came to an end in 1877.

In 1829 Holland rode to Caernarfon to collect wage money from the Williams Bank. He stopped one day at the Pen y Groes Inn, and here in the parlour he met an Irishman, Mr Henry Archer. Archer said he was interested in railways and Holland told him he should take an interest in creating a railway from the Ffestiniog quarries to Porthmadog. He arrived at Plas Penrhyn on 31 December 1829. He stayed there for some days and discussed the prospects for a railway. Holland writes:

He went over the proposed line himself and thought it feasible. He said it must be a single line of about 2 feet wide that it would cost much less, and less to pay for land taken. One evening when he had returned to Plas y Penrhyn he said he had been looking about the quarries, and had arranged with a gentleman who was conversant with engineering and laying out lines and had partly engaged with him to meet him (Mr Archer) to walk over and lay out the line with him – on asking who he was, he said it was Mr Benjamin Smith. I told him that would never do. He asked why – I gave him my reasons and said if any man in the neighbourhood was to be employed, it must be Mr Jas. Spooner (the father of Mr Chas. Spooner). He (Archer) asked why, was he an Engineer, could he do it, etc. I gave him my reasons, said he could do it, etc. We arranged to go and see him the next day. He was then living at Tanyrallt Isa near Tremadoc. This was the 16 Jan 1830, we called upon Spooner, who was in very low spirits having so lately lost his eldest daughter. I explained to Mr Spooner the object of our calling and that Mr Archer was ready to arrange with him, to lay out the proposed line. He declined to undertake it – said he was not a sufficient engineer to undertake it. I told him, he must undertake it and that he might get a man who had been under Robert Stevenson, laying out the line between Chester and Holyhead and that as the line was nearly finished, there was a chance of getting a good man. He consequently arranged to go off the next day to Anglesey to enquire for a man. He met with one of Stevenson's head men and learnt from him that there was a very good & competent man that he could spare – one Thomas Pritchard. Mr Spooner saw him and agreed with him to come over which they did and they – Archer, Spooner and Pritchard (with two of Spooner's sons) set to work and accomplished the marking and mapping of it out, and according to that marking out, it was eventually made. The company of Mr Archer then went to Parliament to get a Bill for the making of the line ... the Bill was thrown out in consequence of the word 'Heirs' having been left out in the contract ... A more regular and genuine Company was formed; many of Archer's Dublin and Irish friends joined and took shares, so that all and everything was on a better footing before Parliament assembled the next year. The Bill was carried and the making of the line was immediately commenced – the first contract for a considerable length was let to a man named Smith from Caernarvon. Mr W. G. Oakeley laid the first stone near Creua Febr 26, 1833. The work of the railway went on and were eventually completed in 1836; we had a grand opening day which was April 20, 1836.

There was great cheering and rock cannon firing all along the line and on our arrival in Port Madoc over the embankment we were drawn by horses; the horses [to descend] rode at the tail of the train, in boxes made purposely, feeding all the way. At Port Madoc we were met by crowds of people, bands playing and the workmen had a good dinner given them. All the better company were entertained at Morfa Lodge by Mr Spooner and ended the day with a dance etc ... I used the railway for two years for carrying my slates before the other companies came upon it, having fixed upon a wharf at Port Madoc which had been made at Mr Madock's (or the estate's) expense and rented it for the remainder of my time, say till the end of the year 1877 at £10 per annum. All the other companies after agreeing to use the railway had to make wharves at their own expense and also pay the Madocks estate a high rent for them.

Eventually there were ten slate quarries being worked and all their slates were brought down by train to Port Madoc; the traffic on the railway increased so much that horses could not do the work required that it would be necessary to have locomotive power. Mr Spooner made out plans of the line, showing the gradients, curves, etc, height of two tunnels etc. these he sent and some that he gave me, I sent to any locomotive engine builders that we could hear of and all gave an answer that no engine could be built to work so narrow a gauge and round such short curves – the late Mr Robert Stevenson and Mr Rastrick, both eminent engineers were had down [sic] (though not at the same time) and Archer, Spooner and myself and one or two others (Thomas Pritchard of course) walked up and down the line with them, they took particular notice of the curves as well as the gradient and each declared that no locomotive engine could be planned to work on such a line. My nephew, Charles Menzies Holland, who was staying with me and was also engaged in making a line between the terminus of the Festiniog line at Blaenau, to Festiniog village and who was brought up as a mechanical engineer used often to go up and down the line of the little railway with me, and frequently expressed an opinion that engines could be built to work up and down the Festiniog line. He consequently drew many plans and wrote a pamphlet on the subject of locomotives. One day he told me, that he was certain he had planned such a locomotive as would work up and down the Festiniog line. He expressed himself so confidently that I mailed it to Mr Spooner and requested him to call the committee of the company together that my nephew may have an opportunity of explaining his ideas, plans etc. Such committee was called. He attended it and explained his ideas very fully and showed them. One of the committee who understood a little of engineering more than the others, thought Mr Holland's ideas good and advised the committee to order engines and arranged with C. Holland to superintend the building of them, if they found an engine builder who would build such. They communicated with a Mr England, an engine builder in London and asked C. Holland to call upon him and communicate his ideas – which was done and consequently Mr England undertook to build two engines according to C. Holland's plans or ideas ... After they were completed, Messers England sent them down by rail to Caernarvon and they were brought by wagons along the turnpike road to Port Madoc and as soon as could be, were put upon the little quarry line – one engine being driven by Mr England, the other by Mr Chas. Holland, they were called the 'Lord Palmerston' out of compliment to his lordship who was then the principal proprietor of the Welsh Slate Company, the other was called 'The Mountaineer'. Both engines worked admirably, the first go off. There was a large attendance of company to see them work as well as quarrymen and country people: these being the first locomotives in the quarry neighbourhood. They proved very successful. Other locos were ordered and built, but on a larger scale. If the engines had not been invented, the working of the trains, with horses, could not have been carried on.

This account is a valuable historical record, written by a man who was there at the time and who was at the centre of planning and investment in quarries and in the railway. However, it seems that his memory is at fault in one respect. Charles E. Lee, in his book *Narrow-Gauge Railways in North Wales* (p. 108), refers to,

epoch-making engines ... the first pair were No 1 *The Princess* and No 2 *The Prince*, both built in 1863 by George England and Company. They were 0-4-0 side tanks, with a separate tender for the coal-space and tool box. The wheels were 2 ft in diameter, the wheel-base 5 feet, the cylinders 8 inch by 12 inch, the boiler pressure 200 lbs and her weight in working order 7 tons 10 cwt. The tank sides came above the level of the boiler tops, ornamental bell-mouths adorned the chimneys, and the domes, which had the whistles on top, had round-topped covers of polished brass ... The nameplates of polished brass were fixed in the middle of the side tanks, with the Prince of Wales's feathers painted over them and the numbers were similarly painted over the builders' plates on the footplate sidesheets. These engines, which cost about £900 each, went into service in June, 1863, and were joined shortly afterwards by No 3 *Mountaineer*, and No 4 *Palmerston*.

These last two were designed by Charles Menzies Holland.

The Festipedia section of the website of the Festiniog Railway Society contains a section on steam locomotives built for the Festiniog railway, and we can assume that they have access to the best information. Here the first engines (all built by George England & Co.) are named in historical order: *The Princess, The Prince, Mountaineer, Palmerston, Welsh Pony, Little Giant*.

The periodical *Engineering* of 4 October 1867 carried a letter from Charles M. Holland (of Tan-y-Bwlch) which said: 'Your recent article ... states that "Mr England was employed under Mr Spooner to design and construct an engine," Will you kindly correct this, as the engines were constructed by Mr England, under my supervision and control, and according to my design? I also advocated the original adoption of the locomotives on the 2ft 1in gauge.' This confirms the words of

Samuel Holland in regard to his nephew and his role in designing the first engines and supervising their manufacture, but the reference to the gauge is doubtful given that Charles Menzies cites the wrong line width, which was actually less than 2 feet.

Undoubtedly, in this whole account, the importance given to James Spooner, Henry Archer and Charles M. Holland as prime movers in the history of the Festiniog railway, along with Samuel Holland and James Pritchard (although we do not know enough about him, but the 'of course' in the above extract is significant) clearly identify the genuine pioneers.

John Whitehead Greaves

In 1833 John W. Greaves, with his friend Mr Shelton, purchased the Llechwedd Quarry, Blaenau Ffestiniog. He came from Leamington and was the son of a local banker. He lived in TanyrAllt isaf, Tremadog, and he married Miss Steadman of Ty Nanney, Tremadog. After the death of the occupant, Dr Carreg moved to live in Madocks's old house, TanyrAllt Uchaf. With the success of the quarry, Llechwedd, he built himself a mansion nearby. He died in Brighton, where he had moved to because of his health on 12 February 1880, aged seventy-three: he is buried in Leamington Cemetery (see *Gestiana* p. 88).

Greaves was committed to public education at a time when schools relied on public support. He supported the new national school in Snowdon Street, Porthmadog, and paid the salaries of schoolteachers. In the 1850s the Llechwedd Quarry was one of the most successful and well-run quarries in the area, with much of this attributed to John Thomas, clerk of the quarry, under the name J. W. Greaves & Sons. John Whitehead's eldest son, J. E. Greaves of Broneifion, was Lord Lieutenant of Caernarfonshire in the 1850s. John's daughter, Evelyn, was married to Osmond Williams of Deudraeth Castle, Minffordd, and had a ship named after her, the brig *Evelyn*, built at Porthmadog in 1877. The brig regularly sailed the east coast of South America, including the Welsh colony port of Port Mdryn in Patagonia.

7

The Spooners of Porthmadog

The originator, James Spooner, was a man of Worcestershire born in 1789. He was a surveyor of land. After his marriage to Mary (from a wealthy Birmingham family) in 1813 they lived in Hafod Tan y Graig, Beddgelert, where Charles Easton was born. In 1818 they lived at Glanwilliam, Maentwrog. In 1825 the family took a lease on Madocks's house Tan-yr-Allt. They followed the deceased Madocks again in 1829 when they moved to the house he bought, altered but never lived in, Morfa Lodge in Porthmadog.

In 1830 Henry Archer commissioned James Spooner to survey a suitable route for the Festiniog railway, from Porthmadog to Blaenau Ffestiniog. Spooner had close help from his two sons James Swinton and Charles Easton as well as from Thomas Pritchard, an Anglesey man born in 1789, who worked for George Stephenson in building the Liverpool and Manchester railway.

Stephenson's son Robert assisted in establishing the route: he walked it along with Pritchard, Archer, Spooner and his sons. Pritchard was employed directly by the Festiniog railway from 1834 and stayed with the company until he died in the 1860s.

James Spooner is said to have come to the area on a pleasure excursion to hunt rabbits in the area where Porthmadog later stood. He is said to have taken an interest in surveying while in that area he met ordnance surveyors at work. He returned to help Madocks with his work.

In 1832 James Spooner gathered an influential local committee to originate the Festiniog Railway Company. Parliament was approached and on 23 May 1832 the Royal Assent was granted for the Act (2 & 3 Will. 1V cap 48) incorporating the Festiniog Railway Company. This was granted powers to raise capital of £24,185 and loans of £10,000 for the purpose of constructing a railway or tram road from 'Portmadoc' to various slate quarries. The wording of the Act included:

> ... will be the means of opening a more direct, easy, cheap and commodious communication between the interior of the principal district of slate and other quarries in the County of merioneth and the various shiping places ... and will greatly facilitate the conveyance of coals and other heavy articles to the several slate and other quarries and mines in the said district, and the conveyance of slates, copper and other ores ... to the seaside.

The Caernarfon Herald reported,

> The Bill which Mr Archer has been indefatigable in procuring for the purpose was passed on Tuesday. That gentleman has entitled himself to the gratitude and respect of the neighbourhood and indeed of the country and especially of the industrious poor.

The first foundation stone of the new line was laid on 26 February 1833 In May 1834 the Festiniog Railway Company started constructing the line by use of direct labour and we can guess that Pritchard played an important part. The contractor was James Smith (who was later dismissed). Astonishingly, in a little over three years the line was completed.

The difference in levels between the slate quarries at Blaenau Ffestiniog and the coast at Porthmadog was 700 feet. In some places the line was cut into rock face hundreds of feet above the valley. Deep ravines, some 600 feet deep, were crossed by constructing narrow stone embankments. Cuttings with a depth of over 20 feet had to be cut into solid rock. The original contract included the construction of two tunnels. One tunnel, 60 yards long, 8 feet wide and 9 feet 6 inches high, was driven through slate to the east of Tan-y-Bwlch station. The second intended tunnel was not built but was replaced by an incline running over a spur in the Moalwyn Mountain. The steepest gradient in the line's length is 1 in 79. The railway is a series of curves, following the contour of the hills. It was described as being 12¼ miles long and its average gradient is 1 in 92.

James Spooner's son, Charles Easten Spooner, was fifteen when he left school and started work in the construction of the railway.

James Spooner played a major part in designing all aspects of the railway, which used horses for pulling empty rolling stock uphill. This upwards journey took 4 to 5 hours. A continuous grade of 1 in 80 constituted the downhill route. This horse-drawn first system was heavily used, but its drawback was that it was slow.

James Spooner became clerk to the Festiniog Railway Company. His son Charles worked on the railway with him, becoming treasurer in 1848 and in 1856, after his father's death and at the age of sixty-seven, he became manager. James Spooner is buried in the family vault at Ynys Cynhaiarn church.

Charles Easton Spooner (1818–1889), son of James Spooner. He was a major figure in the building and development of the Festiniog railway.

Charles Easten Spooner dominated all aspects of the railway for the next thirty years, until he passed away in 1887. He picked an experienced and able group of engineers, working under William Williams, Superintendent of Railways, who developed Boston Lodge into a high level locomotive and rolling stock manufacturing centre and repair workshop.

The following appeared in the periodical *The English Mechanic,* dated 20 November 1889:

The death is announced of Mr Charles Easton Spooner, whose name is closely associated with the designing and development of Narrow Gauge Railways. He was born at Glanwilliam, Maentwrog, [sic] in the year 1818. After leaving school he was engaged with his father from 1832 to 1836 in engineering and constructing the Festiniog Railway, which was designed by his father for the purpose of conveying the slate from the quarries in the hills to Portmadoc. The line, which is 13 miles in length, has a gauge of only 1 foot 11 and a half inches and was scooped out of the cliffs overhanging the Vale of Ffestiniog. On his father's death Mr C. E. Spooner became manager and engineer of the line and did much to advance knowledge of the system by his addresses and writing. The Festiniog Railway thus became the pioneer of narrow gauge railways throughout the world and Mr Spooner's 'Narrow Gauge Railways' which was published by Spon in 1871 is well known to all interested in the work of constructing small railways in difficult places.

Steam locomotion did not begin until 1863 on this railway, after the steam method had become more developed. Additionally, heavier steel rails had to be laid before steam was introduced.

The Festiniog Company's locomotives had wheels and tyres cast in one piece. The usual procedure is to create the tyre separately and shrink them into position. But it was found that due to the gradients involved and the extreme application of brakes, the outer section would become displaced, so this method was abandoned in favour of a solid construction.

In 1863 the Festiniog Railway Company ran a steam engine on their 13½-mile narrow gauge track over a mostly virgin upland landscape. When passing loops were constructed, trains kept to the left, a unique feature.

Robert Francis Fairlie was commissioned by Charles Easton Spooner to design and build *Little Wonder*, an articulated locomotive, suitable for the heavily curved and steeply graded track. Trials in 1870 were very successful.

In December 1878, a travelling reporter from the American *Scribner's Monthly* visited the railway and wrote of Charles Easton as 'a hale, dignified gentleman of sixty. His son, Percy, a young man of energetic and companionable traits, my guide in the subsequent explorations, assures me that he is not beyond the ability to turn "cartwheels" with the agility of youth.'

The authority he carried and the respect people had for him in the sphere of Porthmadog's industries caused a new sailing ship to be named after him. It carried a bust in copper of him on its prow. It was the *C. E. Spooner*, a jack barquentine, built at the south end of Cei Newydd by David Jones in 1878. She was 103 feet in length, 24 feet in width and 12 feet in depth (numbers simplified). She crossed the Atlantic in the very fast time of thirteen days, 4 hours from Shoal Bay, Labrador to Fastnet Rock (Ireland) and later sailed the return passage in eighteen days. Charles Easton used to stand on the garden of Bron-y-Garth and wave to the vessel as it passed by.

Charles Easton's son George Percival (Percy) Spooner worked as a designer for the family firm, from 1872 to 1879, and designed a number of their best engines. He designed railway carriages which ran on bogies, for the first time in this country. He left for India where he became locomotive superintendent for the Indian State Railways. He died in 1917 after returning to England.

Charles Easton Spooner. A picture taken in 1869 when the miniature steam train built by William Williams and his engineering team at Boston Lodge was laid on and now in the garden of Spooner's home, Bron-y-Garth. The boy is likely to be Johnny Williams, Williams's son, who went to India in 1886.

Another son of Charles Easton, Charles Edwin, also worked for the railway. He left to work in Ceylon (now Sri Lanka) in 1876 and became general manager of the Federated Malay States Railways from 1901 until his death in May 1909. He was also involved in civil engineering and the design of public buildings. There is a Spooner Road named after him in Kuala Lumpur.

In the house, Bron y Garth, where the Spooner family lived, in 1869 a garden railway was laid out. The brass trackwork, engine and rolling stock were made in the FR works at Boston Lodge (where part of the trackwork is now stored). The miniature steam engine *Topsy* was built by William Williams (foreman of the FR locomotive department) and is the only one of its kind in the world. He built his own tools for the job, and his original brass spirit level, only 4 inches long, still exists. The magnificent *Topsy* can be seen in a glass case in Spooners Café at the railway. It was built on the dimensions of one seventh of an early FR locomotive built in London by George England. It is 1 foot 10 inches in length and ran on a 3.125 gauge. It was loaned to the railway in 1963 by Edwin Williams, Penmaenmawr, William Williams's grandson, who spent his career in India. William Williams's son John, Edwin's father, left Porthmadog for India in 1886 for a position with Percy Spooner; he became a director of the East Bengal Railway Company.

8

Marketing Slate

The quarrying and mining of slate in Blaenau Ffestiniog continued at an increasing rate through the nineteenth century due to the solving of the problem of getting slate out of the mountains, down into the harbour at Port, and away to market. Until the middle of the century, the market was home – Wales and England. The jurors at the Great Exhibition in London of 1851 remarked

> These slates, which are of superior quality, are chiefly in use for covering buildings. They are also employed as walls for cisterns to hold Water, but in this case large slabs are required. Several such slabs are exhibited, some of which are more than 15 feet by 8 feet ... The jury have awarded a prize medal to Mr John W. Greaves.

The Paris Exhibition of 1854 gave Samuel Holland an 'honourable mention'. So, owners and merchants were aiming at a wider market.

Two significant markets were Australia and Germany. Australia had its gold rush in 1851 and its cities (needing slate for roofing) expanded rapidly, including Melbourne. In 1856 Owen Morris of Porthmadog recorded that 'in 1852, 3 and 4 large quantities [of slate] were exported from Porthmadoc to Melbourne, Geelong, etc by way of Liverpool where they were reshipped into the Australian liners.'

Exports to the United States, especially to the Gulf of Mexico ports had started as early as the 1820s, principally by the *Gomer*, under Captain Pritchard. The *Gomer* was built in Minffordd in 1821; she was 73 feet long, 23 feet wide and 23 feet in depth (these numbers have been simplified). She traded as a passenger ship between Beaumaris and New York. She was lost off Great Orme's head, Llandudno, in 1879.

After 1842 local schooners regularly took slate to Hamburg, on the River Elbe in northern Germany. In May 1842 Hamburg experienced a serious fire and 2,000 houses were destroyed. It is reported that Mr Mathews, owner of Rhiwbryfdir Quarry, consequently went to Hamburg and persuaded the city architect to use his slate rather than Penrhyn slate which had previously been selected for roofing new public buildings. The large number of roofs of thatch subject to incineration in Germany added to the market for Welsh slate.

In 1856, about 10,000 tons of slate was exported to foreign countries. Railway stations, churches and public buildings were roofed with them. The Baltic countries – Copenhagen, Stettin, Danzig, etc., – also received Welsh slate.

The total value of exports of Welsh slate to Western Europe rose from £5,000 in 1866 to £226,000 to Germany alone in 1876, according to Hughes and Eames (*Porthmadog Ships*, p. 31).

Porthmadog ships lined up on the Elbe to disembark their slate cargo. A studio picture taken in Hamburg shows ten Porthmadog-based master mariners, all with Welsh names except one – Captain W. Green. The most recently built ship named in the caption is the *M. A. James*, built in 1900, so we can assume that the picture was taken subsequently. The remarkable thing is that all these ten captains are waiting for their ships to be unloaded of Welsh slate in the Elbe: so vigorous was the trade between Wales and Germany. Canals, railways and roads were built or improved throughout Europe and these carried Welsh slate to a wide sector of market.

However, the First World War ruined the Porthmadog slate trade, and all imports through Hamburg ceased in 1914.

The last ship to be built in Porthmadog was the schooner *Gestiana*, in 1913. Four shares from her sixty-four were owned by William George of Garth Celyn, Criccieth, described as a solicitor: he was David Lloyd George's brother, born 1867. The prime minister's wife, Margaret Lloyd George, of Bryn Awelon, Criccieth, owned one share. When the ship was being 'christened' – launched – the glass bottle thrown against her timbers failed to break. This was considered a bad omen. This ship was lost off the coast of Newfoundland on her first voyage during October 1913.

9

Blaenau Ffestiniog: The Main Quarries and Mines

Llan Ffestiniog existed as a community in the eighteenth century (the word *llan* meaning established settlement or village), but Blaenau Ffestiniog did not. The latter settlement grew on the back of the slate extraction industry.

The expert Alun John Richards calls it,

a compact area of abundant Ordovician slate ... some of the largest and most efficient workings in the industry, producing in the latter part of the 19th century almost one third of Welsh output. Due to the steep northward dip of the five great veins, virtually all work was underground ... sited at the last bleak outreach of Meirionnydd, where the Bowydd and the Barlwyd coalesce; glowered over by the Moelwyn and Manod mountains, the quarries form an amphitheatre with at their feet Blaenau Ffestiniog and its outliers Manod and Tanygrisiau.

It was working underground (mining) that characterised slate workings in Blaenau Ffestiniog and which took the business there to heights of employment and profitability.

We have already covered some of the ground in identifying the slate pioneers and their workings. Now we move to the overall picture. Alun John Richards (*Gazeteer of Slate Quarrying in Wales*, p. 176) identifies twenty-seven slate workings in Blaenau Ffestiniog. If we imagine taking the A470 from Betws-y-Coed, along the Lledr valley, through Dolwyddelan, we come to the Crimea, where the road dips and comes into Blaenau Ffestiniog between heaps of slate waste. On the left are the Llechwedd quarry/mineworkings, now a visitor centre including a trip through underground workings that is well worth taking. To the right is Glodfa Ganol, now worked for its stone but not open to the public. This is the site of the original Oakeley quarry, which, like its neighbour Llechwedd, was a mine, not just a quarry.

Entering the main road of the town from the Crimea, on the right is the area of the original Rhiwbryfdir farm, and turning left, at the end of the built-up area, the Bowydd workings are on the left, with Diphwys close by and then Maenofferen.

At the junction of the A470 (Crimea) with the main street, the Festiniog railway comes in through Tanygrisiau and halts adjacent to the mainline railway.

We have identified Cadwalader Jones as the first Blaenau pioneer, going on to employ twenty men, but at the turn of the century their landlord sold the site over their heads to a syndicate of Lakeland men, William Turner and the brothers David and William Casson. Turner had quarried in Ireland and had married a Wicklow girl. The Diphwys quarry moved ahead quickly and secured a valuable contract to roof military barracks. Transport by packhorse or donkey was initially the method of carrying slates to the Ffestiniog Valley and on, by other means to the Dwyryd, but Turner constructed a road to the valley ridge site known as Congl y Wall (the corner of the wall), making drawn wagons the new transport method. He was the first to advocate the use of iron rails, rather than wooden rails sheathed in iron. His wagons ran on wider tracks than the later Festiniog railway gauge, which later caused transport problems.

It was the Casson brothers who brought about a revolution in Blaenau Ffestiniog slate extraction, after Turner's time. The veins of slate were steeply dipping, so it followed that their main bulk was underground, sometimes very deeply underground. The design of underground workings is a study in itself. Digging out holes in a mountain was clearly a difficult procedure, and potentially very dangerous, relying heavily on the advice of those who had worked on the terrain and knew its underground structure. Sometimes as many as twenty holes or caverns were created, some directly underneath others, so the walls between them and the floors underneath had to be stout enough to hold enormous weights.

The Casson brothers had other interests; they established Cassons Bank. In the 1840s it had branches in Porthmadog, Blaenau Ffestiniog and Pwllheli. It was later incorporated into the North and South Wales Bank, later the Midland Bank, now HSBC.

At Blaenau Ffestiniog, slate output had gone up from 500 tons at the beginning of the century to 12,000 tons by the early 1820s. This output, most of it by export, was enlarged by the decision of Boston, USA's city council to insist that roofs of all new buildings should be slated.

Diffwys was the largest working quarry, with Manod, on top of the Manod mountain, close behind. Both quarries were frustrated by the difficulties of getting slate to market by the River Dwyryd route. The next quarry to make an impression was the one owned by father and son Samuel Holland. They had, as we have noted, taken a three-year take note (a form of lease) on Oakeley land at Rhiwbryfdir on the slopes of Cwm Barlwyd. This proved very successful.

In 1823 Lord Newborough's old workings at Bowydd were set up again and were copying Diffwys's underground methods. Improvements came on, including laying rails for underground transport, mechanical slate sawing and mechanical slate block cutting.

Lord Rothschild in 1825 formed the Welsh Slate and Copper Mining Company and offered Holland £28,000 for his Rhiwbryfdir quarry. Only a few years were to run on his lease, so Holland sold up, making what in today's money would be more than one and a half million pounds. Although the new company would later become very successful, for the first fifteen years of its life it steadily lost money and became known for its late payment of wages. It also failed to pay dividends so Rothschild was replaced by Lord Palmerston as chairman of the company (see *Slate Quarrying in Wales* by Alun John Richards, p. 48).

Blaenau Ffestiniog slate dominated slate production in North Wales in the mid-nineteenth century. It started with the Diffwys quarry's 6,000 annual tonnage in the 1820s. The Welsh Slate Company reached 4,000 tons in the 1830s. Samuel Holland's new quarry at Cesail prospered in the 1830s. Nathaniel Mathew successfully used the Oakeley land to create Glothfa Ganol or Middle Quarry.

Holland had improved road communications between his quarry and the Dwyryd at Maentwrog but the river was only useable at mid and high tide. The small boats shipping slates down the Dwyryd, some 20,000 tons a year, had become very inefficient by the mid-1830s. If Holland could find a better way to market, his output would challenge those of the other great Welsh quarries, at Penrhyn, Dinorwig and Nantlle.

The Llechwedd quarry – shortly to become a huge mine – opened in 1846.

In Hamburg in 1842 there was a serious fire, striking the wood-ribbed houses and their combustible straw roofs. Nathaniel Matthew went quickly over to Hamburg to sell his slate to house builders and slate merchants, resulting in the substantial trade in slate made with Germany over the rest of the century. This contributed to the building of suitable merchant vessels in Porthmadog, and mercantile trade with Hamburg via the North Sea and down the Elbe continued vigorously until it ended with the First World War.

In the 1820s, the Diphwys and Bowydd quarries pioneered underground slate working. The working of underground caverns started with the underside of a vein. Caverns were worked downwards, so that material fell downwards. Caverns were in the region of 50 feet wide, with slate pillars in between of 30 feet width. Temperatures were steady, in the region of 55 degrees all year, so at least extreme winter cold was avoided by the underground workers.

Although gas lighting was installed by Holland in 1842 in a 1,000-foot-long tunnel, there was no lighting in most mines other than candlelight. Candles were issued to miners by the owners and the cost deducted from their wages; the cost in the early years amounting to £1 to cover four weeks' supply. Interestingly, at the Oakeley quarry in the 1890s, the widow of a man killed at work asked for her compensation of a mere £10 to be paid to her as a lump sum so that she could buy stock for opening a candle shop. It was not until the 1950s that candles were ousted by the introduction of the Oldham electric cap lamp.

By 1850 more people in Wales were employed in mining and quarrying than were employed in agriculture, thus officially becoming the first industrial nation on earth.

The 1860s saw great expansion in slate production; by 1869 the price of slate was up 25 per cent from ten years earlier and the Welsh Slate Company's output almost doubled in five years. Diphwys Quarry was sold in 1863 for an incredible £120,000, which was 120 times its 1801 cost. Between 1860 and 1870 the quarryman's wages rose from 4s 6d per day, to 5s 6d. (There were 20s to a pound and 12d to a shilling.) Blaenau miners were paid 1s less per day, and a labourer 1s less than that.

From 1840 to 1870 the highest tonnage was achieved by the Welsh Slate Company, at 43,296 tons in 1870. The Llechwedd Quarry produced less than half that, and eight other quarries produced down to Wrysgan's 2,078 tons. In 1882 it was reported that out of fifteen quarries in Blaenau Ffestiniog, only four were profitable.

The population of Ffestiniog parish rose from only a few dozen in 1780 to 11,274 in 1881.

After shortcomings in working practices, the WSC had a serious fall and consequently most of their underground workings were destroyed, as well as damage done to Holland's and Mathew's workings above. W. E. Oakeley took over the quarries which were on his land and by 1900 the great Oakeley Quarry consisted of the previous Holland's, Mathew's, Welsh Slate, Nyth y Gigfran and Cwmeithin. It had twelve mills and 500 saw tables, but it never repeated the profits of the '60s and '70s.

By 1884, slate prices in the market had declined. Interestingly, one reason for this was the importing of slates from the USA, where some North Wales quarrymen had settled and begun working in slate quarries. In 1869, the Welsh language was being spoken in American quarries and roofing slate and writing slates were being produced in high quantities.

By 1894 the industry had picked up again, with a 30 per cent increase in production value over that in 1890. In 1895 an extraordinary 916 men worked above ground and 736 below ground at the Oakeley slate workings (Rhiwbryfdir, Cesail, Gloddfa Ganol) in Blaenau Ffestiniog. This was by far the largest in the area, producing over 56,000 tons. The second largest was Llechwedd with over 15,000 tons output and a payroll of 486 men. Richards lists fifteen working slate quarries in Blaenau (p. 192).

However, the reputation of the Blaenau Ffestiniog mines and quarries never recovered after the publication in 1895 of the *Report of the Enquiry into Merionethshire Mines*. We can do no better than to quote Alun John Richards's masterly summary:

Despite robust rebuttals from medical and management witnesses, the report made clear how unhealthy and hazardous slate quarrying was. It showed that although gas posed no dangers of explosion or asphyxiation, it was more dangerous to work in an underground slate quarry than even the most notorious of coalmines. Roofs crumbling, falling rock, blocks slipping during handling, missed footings in the dark, stumbling near the many sheer drops, running down by wagons – there were a hundred dangers. It was scarcely safer in the open quarries. Wet or greasy rock, and precipitous pits reached by rickety ladders, presented additional hazards. In all workings there were winches and hand cranes whose ratchets slipped, sending winding handles spinning to kill a man. Blasting had been made safer by the obligatory provision of blast shelters, and by firing at fixed times with bugle and flag warnings, but horrific accidents from shot firing were still commonplace. [There was a hospital at] ... Oakeley, other owners at Blaenau Ffestiniog had put the town ahead of many municipalities by sponsoring a hospital in 1848 ... However, even with a hospital on site, it could be a lengthy and painful journey for a patient to reach it from a distant workface. Thus fractures and other injuries not ordinarily life-threatening could result in death, or best, permanent disablement.

The report drew attention to the chronically bad diet of the men, though this was passed off as *'failure of the wives to provide proper meals'*. It highlighted the great incidence of diseases, particularly pulmonary complaints, some 50% of male deaths being from respiratory causes. These, some witnesses attributed, not to dust, damp and bad housing, but to drinking *'stewed tea'*. It would be the 1920s before the canard that slate dust was 'beneficial' was finally laid to rest. Even so, it was the 1930s before dust extraction and dust suppression was seriously tackled, and dust-induced lung diseases were recognized as compensatable complaints, for which employers could be held liable. Since by this time many employers had gone out of business, victims were denied any source of redress. It was not until 1979 that a Government scheme was enacted to grant such men, or such few of them who were then still surviving, appropriate compensation.

The report also drew attention to the awful conditions prevailing in barracks and the plight of men who having to walk many miles in rain to work, would spend the day in wet clothes, either on exposed rock faces, draughty sheds or in watery conditions underground. As a result, serious disability was more likely to be the result of disease than traumatic causes. Added to which, even in the large quarries with sick pay schemes, economic necessity could force an ill or injured man to return to work before recovery was complete. For all its faults the report was a damning condemnation of conditions in the industry. (pp. 182–3, *Slate Quarrying in Wales* by Alun John Richards).

The twentieth century saw a decline in the slate industry. Although slate for roofing continued in production, the 'slab' trade declined. Fashion took against slate for flooring and slate in public lavatories was replaced by other material. Italian slate was imported for use in snooker tables, which was 1 inch thick compared with the Welsh 2 inches. The underlying reason for the decline was the absence of change and modernisation in working practices. The First World War took in men who would otherwise have worked in slate, so output declined owing to a lack of labour.

The 1920s saw an upturn in Blaenau Ffestiniog's slate output, but in the 1930s slate employment was down. There were only eight quarries in North Wales with over 100 men employed, and four of these were in Blaenau Ffestiniog (Oakeley, Llechwedd, Maenofferen, Votty & Bowydd).

Roofing materials were being imported from the Continent, including cheap French slate and cement tiles which were half the cost of Welsh slates. Good rock in the Welsh quarries was becoming more difficult to find and waste shifting was becoming more necessary and was expensive.

In 1970, the biggest quarry in Blaenau Ffestiniog, Oakeley, closed. In the late 1970s only 20,000 tons of finished slate was being produced by the Welsh quarries, and Blaenau Ffestiniog's population in 1970 was half what it was in 1890. Llechwedd, under the Greaves company, continued and opened a successful tourist attraction under the name Quarry Tours; here visitors can enter a descending tramway and have a unique experience of underground workings.

10

Building Ships

According to Emrys Hughes and Aled Eames, in their magisterial volume *Porthmadog Ships*, serious shipbuilding started in the Porthmadog area in 1776. Between then and early 1824, which was before the harbour in Port (as it is always called locally) was functional for shipping, nine sloops and one brigantine were built on the extensive seashore around the inlet knows as Traeth Mawr and on the banks of the River Dwyfor (Traeth Bach) as it came into the estuary. The brigantine *Endeavour* of 87 tons is described as being built in Borth-y-Gest although other references have it built at Traeth Mawr in 1802: she was 66 feet in length with a width of 221 feet (these numbers are simplified). The Hughes/Eames notes say that she was captured by the French in 1808 and that in one long voyage she carried a load of slates from Porthmadog to Archangel. She cost £1,958 3s 2d to build.

Barmouth and Pwllheli built a substantial number of sailing ships, and between them and Porthmadog/Borth-y-Gest, the western shore of central and northern Wales saw the building of over 500 sailing vessels between 1776 and 1913. Shipbuilding in this area really was an industry.

Between 1798 and 1825 the value of slate in the market doubled. In 1825 the new harbour at Porthmadog saw some 12,000 tons of slate carried away onboard ships. However, carriage down from the mountain was a major problem. A new narrow-gauge rail link, over some 12 miles of the most treacherous terrain, was surveyed and in 1836, under the ownership of the Festiniog Railway Company, it was open for business. This followed from some of the most spectacular and back-breaking periods of work by gangs of men working with only hand tools.

Looking at *Porthmadog Ships* by Emrys Hughes and Aled Eames, we are struck by the large amount of detail included in the listing of ships; this list starts on page 146 and finishes on page 275, with an average of about four vessels per page. Emrys Hughes calls it, 'a brief record of vessels trading to and from the port during the greater part of the 18th century and the first quarter of the 20th century with short notes on those I have personally known and heard of.' He even includes the names of shareholders and their holdings out of sixty-four. One of the shortest entries comes on page 271, as follows:

WILLIAM ALEXANDER
Sr [schooner], 89 tn. [tons net]. Built at P.M., 1839, William Jones. P. M. ma [master of vessel]
    Lost with all hands, North Sea, 3.10.1860.

That brief, pared-down account seems to show that Emrys Hughes, unlike his usual practice, could find few details of this ship.

However, if we turn our attention in another direction, towards the old church of Ynyscynhaiarn, which was virtually on an island before Madocks and Willliams's first embankment was built, we find a gravestone inscribed:

Sacred to the memory of William, son of John Jones, Schooner 'William Alexander' of Portmadoc, by Margaret, his wife, died October 21st, 1857, aged 5 years. Also the above-named John Jones, aged 49 years, and his eldest son, John, aged 16 years, who with their vessel and all hands were lost in the North Sea in the memorable and awful gale on 3rd October 1860. [A poem by Emrys, translated from Welsh, is inscribed:]

Oh, loss of three; Oh, the blow
to the wounded feelings,
For dear sons, and the father,
Long will be tears of loss.

In memory of the blameless, separated
Between earth and the deep,
From land and sea will come the dead,
To the same court will happily return.
(from *Gestiana* by Alltud Eifion, p. 125)

Types of Vessels Built

Shipbuilding in Porthmadog (including Borth-y-Gest) is associated with the class of sailing vessel known as schooners. The other main types of sailing ships built here in the eighty years or so before 1913 were: brig, brigantine, barque and barquentine.

A schooner is a ship with two or more masts of near equal height: three masts were introduced later. Sails are essentially of two types: fore and aft, which are large pieces of canvas, hung in parallel with the alignment of the vessel; and square sails, which are rarely square but rectangular. Square sails are hung on a mast across the line of the ship, each one adjustable to catch the wind; three or four on a mast are common but there can be as many as seven.

The simplest type of schooner is the fore and aft schooner: here there are fore and aft sails hung on gaff spars attached at the bottom to a boom. A staysail would be attached to the foremast and stretched forward to attach to the bowsprit. Where there were three masts, they were named foremast, mainmast, mizzen (mast). Where two sails were attached to the bowsprit, the forward one was the jib sail and the one behind this one was the foremast staysail (these are very recognisable because of their long narrow angle at the top).

The 1820s and 1830s in Britain was a period of industrial depression, but in the Blaenau–Porthmadog area industrial activity was picking up, helped by the opening of Porthmadog's harbour in 1824. Export of slate encouraged shipbuilding and suitable shipbuilding sites in Porthmadog and Borth-y-Gest came into use. The basic schooner type was developed and the building of the topsail schooner became the norm. These were designed to carry slate quickly over deep water, particularly the North Sea, for the German trade and the Atlantic (called the 'Western Ocean' by local sailors) for exports to North America. According to Owen F. G. Kilgour (*Caernarfonshire Sail*), 483 schooners were built on the shores of Caernarfonshire between 1800 and 1910, peaking between 1830 and 1870.

The two-masted topsail schooner was the first type to be built during this period, with the three-masted schooner coming later in the 1890s.

The schooner was built mainly in Porthmadog, but the other Caernarfonshire maritime centres also built them, including Pwllheli, Nefyn, Caernarfon, Porth Dinllaen and Borth-y-Gest.

The two-masted topsail schooner had an average dimension of 76 feet in length, 20 feet in width and 11 feet in depth. The average tonnage was 96 tons.

The Caernarfonshire schooner was built of pine planking, carvel-laid (edge-to-edge planking, distinct from clinker-built, which had overlapping planking) on a skeleton framework of oak. The pine was largely imported, the oak being locally sourced when possible. The area of some 10 miles inland from Porthmadog, notably the heavily wooded Vale of Ffestiniog, is now significantly short of oak trees, a legacy of the shipbuilding days.

The hull shape of the schooner evolved over the decades. It was designed to cut through and sail over the rough seas of the Atlantic. It was shallow in the water, comparatively slim, including square sails set high to catch any wind available and it sliced into and used the top of the water. Its bows were wide, narrowing at the stern and the hull had a curve which raised the bow, allowing it to ride over rough seas.

The foremast and mainmast were of Douglas fir. Each was commonly in two parts, the lower mast and the topmast. Both complete masts were of similar height, with the mainmast slightly taller. The bowsprit was large and distinctively raised, pointing well above the horizon. Below the bowsprit, some vessels had an effigy carved out of wood, as did the *Pride of Wales* and the *C. E. Spooner*.

The single topsail schooner characteristically carried nine sails, including four headsails (two attached to the bowsprit), with two large fore and aft sails hung on a gaff and bottomed by a boom, and one square sail on the head of the foremast.

Borth-y-Gest. This is where the *Pride of Wales* was built in 1870 by Simon Jones. She traded between Rangoun and Ceylon (now Sri Lanka) and Surabaya. She grounded in the north Atlantic in a force 12 gale. She was 125 feet long.

The double topsail schooner differed by having two square sails on the head of the foremast, on the topmast.

The two-masted schooner was the early ship of choice. It moved swiftly and was widely used to carry perishable fruit from Portugal and Spain to Britain. It also carried salt codfish or dried fish from Newfoundland across the North Atlantic, a voyage which in one schooner was alleged to have taken ten days. The barquentine *C. E. Spooner* (see Hughes and Eames, p. 168), built in 1878 by David Jones at Porthmadog harbour sailed this route in thirteen days, with a crew of six men. She was the last vessel to deliver salt cod to Liverpool from Newfoundland.

The three-masted topsail schooner was built in Porthmadog from 1891 to 1913. The ill-fated *Gestiana* was one such. It had two fore and aft sails on its mainmast and mizzen, with three square sails at the head of its foremast, in addition to four jib or staysails.

There were smaller and larger types of three-masted topsail schooners. Rounded out, the typical dimensions of the larger type were: 100 feet in length, 22 feet in width and 12 feet in depth. These were long, narrow vessels.

Porthmadog-built vessels, especially the later ones, had a distinguishing feature, the curve of the hull's upper edge, which ran from prow (or stem) to stern, along the gunwale. It gave them a graceful appearance.

The other types of vessels built in Porthmadog/Borth-y-Gest were brigs, brigantines, barques and barquentines. Although hung with fore and aft sails, these types of vessels relied on square sails to give them choice of direction and power.

The brigantine was slightly bigger than the large schooners, with two masts, and carried more freight than the usual three-masted schooner. It was longer, at an average of 93 feet, with a depth of 12 feet. It carried four or five square sails on its foremast, out of its thirteen sails.

The *Blanche Currie* brig, built in Borth-y-Gest in 1875. Lost with all hands in 1914. (Courtesy of Gwynedd Archives)

Borth-y-Gest, the Porthmadog side of the bay. This narrow strip lay where the *Blanche Currie* was built by Richard Jones, Garreg Wen, in 1875. The row of houses here ('Cilan' to the right) were built in 1938.

The barquentine was a big cargo carrier, with an average tonnage of 228 tons. Dimensions, rounded out, were 110 feet in length, 25 feet in width and 14 feet in depth. It had three masts; foremast, mainmast and mizzen. It typically carried fifteen sails so carried more than a schooner.

The jack barquentine had gaff sails (fore and aft) on the main and mizzen masts, with three square sails on its foremast. A gaff is the wooden pole, slanted, that sits across the top of a main sail, attached at one end to a mast.

In Porthmadog, the building of jack barquentines took place mainly in the mid-1870s. They came before the building of three-masted schooners.

A brig is a two-masted mostly square-rigged vessel. It was designed for heavy cargo carrying and did not possess the attractive lines of the schooner. They were flat-bottomed and could be beached without toppling over. The *Mary Holland* was built as a brig in Porthmadog in 1843 and in 1854 converted to a barque. The oddly named snow was an earlier version of the brig.

The barque was a vessel with three or more masts, square-rigged, except the mizzen, which carried a fore and aft sail. They were long, with an average length of 120 feet and were boxy cargo carriers. Captain James Cook's *Endeavour* was a bark (or barque), which sailed around the world in deep water.

11

Porthmadog Shipbuilders

Henry Hughes, (*Immortal Sails*, p. 49) the chronicler of the history of Porthmadog ships and rare chronicler of the working life of a mariner, gives historical pride of place to the snow (or sailing brig) *Gomer*, built on Traeth Bach, Minffordd, in 1821. Her dimensions are given as: 75 feet in length, 20-foot beam and 11 feet 6 inches in depth. Compared with vessels which were built soon after in Porthmadog she was a small ship, and not comfortable or easy for sea-crossing passengers. Perhaps as many as seventy passengers were carried, in addition to her crew of seven or eight hands. Registered in Beaumaris, of 80-ton capacity, she had a distinguished career as an Atlantic passenger vessel, crossing the North Atlantic from Beaumaris to New York. Emigrants travelled from all parts of Wales to join Captain Richard Pritchard on his own ship, the *Gomer*, many of them settling in New York state and in Wisconsin. Captain Pritchard retired and went on to run the National and Provincial Bank branch at his home in Lombard Street, Porthmadog. He died of a fever in the North Pacific in 1855 when he was seventy-two years old.

Henry Hughes's favourite Porthmadog-built sailing vessel was the *Elizabeth*, which was the fourth locally built ship to carry this name. The first was a schooner built at Traeth Mawr (the Glaslyn estuary) in 1804 by Ellis Roberts (see *Porthmadog Ships*, p. 181). This fourth *Elizabeth* during the war of 1914–18 served as a 'Q Ship', a decoy for German U-boats. Hughes writes,

> For over 35 years she sailed about the world with serenity and ease and with little waste of time. True, like most ships, she struck bad patches on occasions, but always came back to make up for it. She and the inimitable *Blodwen* were launched in the early nineties and could be termed contemporaries and rivals. Their building slips were only 50 yards apart. The builders, David Williams, in the case of the *Elizabeth*, and David Jones, in the case of the *Blodwen*, vied with one another in friendly 'scraps' to put the best possible material and workmanship into their respective craft. Both vessels turned out to be little beauties and exceedingly speedy. (*Immortal Sails*, p. 195)

The *Elizabeth* was unusual in that she spent time trading in the Indian Ocean: she regularly called in to Madagascar, the Seychelles, Mauritius and Réunion Island. In 1927 she was hit by a great typhoon when she was heading for Mahe in the Seychelles; she was left as a wreck in a coconut plantation.

Samuel Holland's wharf was the first of any scale to be established at the new port. It was followed by wharves built and owned by the Welsh Slate Company, the Rhiwbryfdir Slate Company and J. W. Greaves, who had to pay rent to the Madocks estate. At Canol-y-Clwt, Henry Jones and his son William built at least thirty sloops – a vessel with one mast and fore and aft sail, with bowsprit. By 1930 they were building larger ships, including the *Lord Palmerston* (111 gross tons) built in 1828. Henry Jones went on to build the barque *Ann Mellhuish* in 1849 and the *Henry Jones* in 1850: both had a gross tonnage of 300 tons.

As a consequence of the increased value of slate and the dire need to get it to market, the year 1826 was a watershed in Porthmadog shipbuilding. Evan Evans and Francis Roberts built sloops and schooners and at Borth-y-Gest shipbuilding started with Robert Owen and William Griffith. Henry Jones was ahead of the game and in 1843 built what was then the largest vessel built in Porthmadog, the snow (later a brig) *Mary Holland*: she was 85 feet long, with a width of 22 feet and a water depth of 14 feet (these dimensions are simplified). She was shorter and wider than the later style ships. She was lost off the Bahamas in 1858.

Henry Jones continued his production at Canol-y-Clwt until 1858 with the creation of the schooner *Ebenezer*. Through his shipbuilding enterprise he had built over sixty ships. His son William was a wealthy man who provided mortgages to potential share owners in the second half of the century. When Henry Jones died, his body and funeral party left Porthmadog by small boats, sailed across the bay to Harlech, where he was buried at Capel Ucha Baptist Chapel.

Other shipbuilders were at work. Evan Evans built the schooner *Humility*. An unusual location, Cei Cwmeithin, which was located on the shore at Trwyn Cae Iago, on the Porthmadog side of Borth-y-Gest, was used by Robert Williams and Walter Williams for building the schooners *Picton*, *Patriot*, *Planet* and the two ships named *Pilgrim*.

The Australia Inn, in Porthmadog's high street, was the home of Daniel Griffith, the shipbuilder. He built the *Volunteer, New Blessing, Sarah, Samuel Holland, Anne Holland, Mary Casson, G&W Jones*, all of them, apart from the brigantine *New Blessing*, were schooners of between 110 and 150 tons.

Between 1846 and 1864, a schooner a year was built by William Griffith at Borth-y-Gest, including the brigantines *Wave* and *Edward Windus*.

A shipbuilding site took up room in part of the area where the Festiniog railway terminus in Porthmadog is now. It was used by Ebenezer Roberts in building the brig *Excelsior* in 1868. She made many voyages to Aruba, carrying away phosphate rock: she had a 'remarkable trading record' (Hughes and Eames). Roberts launched his ships on the south side of Rotten Tare, where the river Dwyryd has created a deep channel.

The *Lord Palmerston* (1828 – lost with all hands in 1869) is described as having a square stern. This is a clue to how Porthmadog ships were beginning to have their own style. Typically, later, they had a distinctive stern: squared off at the back, with the horizontal planking stretching some 6 feet and more downwards. Below that, towards and under the waterline, the stern of these ships had a distinctive sculpting, a smooth half-moon shape. At the other end, and most local ships rarely exceeded 110 feet in length, the prow had a gradual upward curve, pointed, with a long and substantial bowsprit, which gave them presence and elegance in the water. Below that, towards and under the waterline, the front edge curved downwards and backwards, not being near vertical as was usual. Henry Jones built the brig *Mary Holland* in 1843; a ship painter of Trieste captured her, clearly showing the typical Porthmadog features on the prow and stern.

Owen Morris wrote in 1856,

Great improvements have been wrought in the shape of vessels built since so recent a period as thirty years ago. Short, bluff bowed, stout sterned and tub like vessels of small size were those in vogue then. Provided a vessel had good stowage room all was considered well, no regard being apparently paid to her sailing qualities. But gradually a more easy wedge-like overhanging shape was substituted for the bluff perpendicular bow, and a more gradual curvature was adopted for the stern – greater length given – and altogether vessels were modelled with greater regard to sailing and weather qualities than heretofore – stowage not being deemed the only essential quality constituting a good vessel. (*Portmadoc and its Resources*, p. 54)

Porthmadog ships were built to carry slate, which is heavy and which functioned as ballast. Later in the century, Porthmadog ships were built for speed in the water – they were comparatively short and shallow, with their usual depth being in the region of only 10 or 11 feet. They had tall masts and a large spread of canvas. A glance at the proportions of the average schooner, such as the *M. A. James*, built in 1900, shows a depth of only 10 feet 6 inches, a width of 22.7 feet and a length of 89.6 feet.

We refer to Porthmadog ships but this includes those built in Borth-y-Gest. Today this village, on the shoreline a mile or so from Porthmadog in the Criccieth direction, is a quiet place much admired by visitors. Very few visitors or residents know that in the second half of the nineteenth century this inlet featured three shipbuilding yards where some of the largest and finest Porthmadog sailing ships were manufactured. Older residents will tell you of how the ships' captains retired to Borth and sometimes flew their ships' flags from masts in their gardens, or kept the wheel as a memento. Ralph Street, high on the hill overlooking the bay, was built early and originally housed mariners. John Owen, who lived in 38 Ralph Street, Borth-y-Gest, in 1896, owned sixteen of the sixty-four shares in the schooner *Cariad*, which was based at Porthmadog but was built at Salcombe. John Owen, her captain, told Emrys Hughes that the ship had crossed from Eurges in Newfoundland to Oporto in ten days, which is a record for crossing the North Atlantic by sail.

Extraordinarily, this tiny Borth-y-Gest bay was the scene of work by three ship designers who had their own building sites. The outstanding Simon Jones built his ships on what is now the car park. The spectacular *Pride of Wales*, a barque of 299 gross tonnage and 125 feet long, was built here in 1870. She spent much of her time on the Indian Ocean, chartered by the Indian government. A second site for shipbuilding was immediately ahead of him, in front of Tai Pilots. Robert Owen and his nephew William Griffith, who had previously been a farm labourer, built their ships here. Morris Owen also built here, for example he built the barque *Snowdonia* in 1874. She was very large at 419 gross tons, and had dimensions of 138 feet long and 17.2 feet depth. She was lost with all hands off Berwick-upon-Tweed, loaded with phosphate rock. Her Captain J. Roberts's body was brought back to Porthmadog for burial. The local poet Ioan Madog (John Williams, blacksmith) wrote a quatrain in Welsh on seeing a ship bearing his body enter Port's harbour with an effigy of a woman on its prow. Translated into English, this reads:

A statue above the water
A seductive shape for sailors
I asked as an onlooker
From where came the old whore.

A third site at Borth-y-Gest was on the other side (the Porthmadog side) of the bay, directly below the row of detached houses designed, with distinction, by Porthmadog architect David Morris,

Borth-y-Gest, *c.* 1900. The scaffolding in the foreground marks the site where Richard Jones built the brigs *Fleetwing* and *Blanche Currie*. Simon Jones's yard was where the car park is now. The barque, *Snowdonia*, was built in front of Tai Pilots (tall houses to the left). (Courtesy of Gwynedd Archvies)

in 1938. These houses have a long view of the estuary of the two rivers and of the line where it meets the seawaters of Cardigan Bay. On this narrow strip of shoreline, Richard Jones, son of local smallholders of Garreg Wen and entirely self-taught, built his ships, including the *Blanche Currie*, a brig, later a brigantine, in 1875. She was originally built for the trade in phosphate rock, which was carried to Europe as fertiliser from the West Indies.

One of the most remarkable qualities in the formation and working of the port of Porthmadog is how the community pulled together in a determination to create a local maritime business which would be of direct benefit to its citizens. It was cooperative and it saw dozens of men working in different trades (sail-makers, blacksmiths, joiners, etc.) all interconnected in their focus on building new ships and getting slates to market, to the benefit of themselves and their community. One was Hugh Jones of 14 Cornhill, Portmadoc, 'block, pump & spar maker, steering wheels made to order and repaired'. The community had its own bank and its own insurance company. In the 1860s the Nonconformists established a new school and one of the men who made donations towards its cost was Captain G. Griffiths, secretary of the Mutual Marine Insurance Society. This society was founded in Porthmadog in 1840. Emrys Hughes wrote,

Maenofferen slate wharf, Porthmadog, *c.* 1896. The ship *Excelsior* is in the foreground. (Courtesy of Gwynedd Archives)

The deed is based upon an agreement reached at a meeting held at Portmadoc in July 1840. There is as yet no trace of this agreement. If this agreement could be found, the exact nature of its foundation could be ascertained. One thing is obvious, that Barnes, Carreg, Casson, Holland, represented the financial interests. The other signatories covenant with these to form the Society. There is, therefore, no evidence whatsoever to show who were the first movers in this matter. I think it would be very rash to assume that the 300 and odd would approach Holland or anyone else to ask them to evolve this scheme of insurance.

What is clear from this and other sources is the important part this society played in the maritime history of the port. It must be said that the dark side of the activities was the number of ships that foundered and the number of sailors who drowned. All the cemeteries in the vicinity of Porthmadog include graves of deceased sailors, lost at sea. Each finished ship in the port was insured locally, and when a ship was damaged or lost compensation was paid. Little is known about compensation paid to the families of drowned or injured sailors. The Mariner's Cemetery, tucked away in a private place off a narrow entrance on the Criccieth road out of Porthmadog, speaks volumes for this. Nobody has ventured to guess how many local men drowned at sea, but it must be many hundreds.

12

Ships' Insurance

In 1841 the Portmadoc Mutual Ship Insurance Society was set up, with financial and administrative help from John W. Greaves of Tan-yr-Allt and Samuel Holland of Plasypenrhyn. It was the first society of its kind to be formed in North Wales. It had the intention of charging shipowners at lower rates than those offered by the London insurance societies. 320 persons were involved, and contributed financially: they subscribed to the 'Society of persons, interested in ships and vessels belonging to the harbour of Portmadoc, employed in the coasting trade, upon the principle of mutually bearing one onother's burdens ... would be beneficial to the shipping interest, navigation and commerce of the harbour of Portmadoc.' This was an ambitious and bold move and it did more than insurance. It delegated experienced men to examine vessels for seaworthiness and made sure that they were commanded by competent masters. This anticipated the Merchant Shipping Act of 1854 which attempted to establish national standards of maritime competence. Premiums charged were in the region of 2 per cent, compared with 6 per cent by the London insurers and in 1856 it was reported that ships and goods to the value of £100,000 were insured by this local society (see *Portmadoc and its Resources*, p. 55). It was a remarkable undertaking and achievement. Again, typically, it was an example of self-help, of people pulling together for the general good.

The society was well run and at a meeting in the town hall in the January of 1865 it was reported that the value of vessels insured by the society was about £130,000, but that in the preceding year nine vessels had been lost and twenty-four were damaged. However, eleven new vessels had been received into the society, 'and those of a larger and more valuable class than those lost'. Also introduced at that meeting, by Samuel Holland, were new rules governing the loading and unloading of cargo in winter in Gibralter, the Mediterranean, the coast of Africa, the Madeiras, Canaries and Cape Verde islands. This shows, at the very least, how international Porthmadog shipping had become. In April 1865, 153 vessels were listed as insured by this society and in 1867 Holland reported that the value of insurance was now £162,000 – an increase of over £30,000 in two years. In 1869 the insurance value had risen to £243,240. In 1870 nine vessels were lost, and the society paid out £7,450, that is £827.00 per vessel. Unlike other marine insurance companies, the Porthmadog unit was persistently local, asserting in 1880 that 'no ship or vessel should be insured with the Society unless the owner or some part owner thereof interested therein to the extent of one fourth part or upwards shall reside within 15 miles of Portmadoc.' In 1866 a second marine insurance society was formed in Porthmadog, this time to insure against damage to other vessels other than the main insured one – third-party risks.

13

Shipbuilding Expands

In 1856, Owen Morris, the local diarist, wrote that the shipbuilding industry 'employs as many as 100 carpenters, joiners and smiths, and turns out from £20,000 to £25,000 worth of shipping property annually.' He refers to,

> the class generally built here are those made to carry from 120 to 160 tons, none below 70 tons, but a few above 160 tons, this range being found the most suitable one to meet the requirements of the slate trade. Occasionally vessels of 300 burthen have been launched for Liverpool firms to be employed in the South American, Californian and Australian trades. Several ships have been lengthened here – most of them by putting a new part in amidships. This has not only proved favourable to their sailing powers, but profitable in the shape of larger dividends. The bulk of the timber used is imported from Southampton, Colchester, and Gloster [sic]. The remainder is obtained from the country around, and is far preferable on account of its hardness and toughness. There are six shipbuilders now connected with the place, and this branch of trade bids fair to become the source of increased prosperity. (p. 54)

There were 30,000 tons of timber imported to Porthmadog in 1855, including deal, oak and birch from North America, roughly-hewn oak for shipbuilding from Gloucester – the product of the Forest of Dean – and from Southampton and Portsmouth, the produce of the New Forest, Hampshire. This itinerary shows that locally sourced timber was not sufficient for purpose and expanding shipbuilding demanded imported timber.

The most productive years in the history of Porthmadog/Borth-y-Gest shipbuilding were 1877 and 1878. Eighteen ships were built – brigs, schooners and barquentines. The longest was the *Martha Percival*, a barquentine, which was 120 feet long with a width of 27 feet, built by Ebenezer Roberts. The *Frau Minna Petersen*, a much-admired three-masted schooner, was built by Simon Jones in Porthmadog in 1878: she had a tonnage of 165, a length of 102 feet, width of 24 feet and a depth of 12 feet. This established the new style for shorter, lighter ships which were very fast in the water. One voyage, very unusually, saw her sail 125 miles up the delta of the Danube to visit the ports of Galatz and Braila in Romania.

A serious break in shipbuilding came after the building of the jack barquentine, the *C. E. Spooner* by David Jones in 1878. This followed forty years of regular shipbuilding. The *Tony Krogmann* was the last of the square-riggers to be built. Between 1874 and 1878, shipments of slate by sea fell, while shipments by rail had increased. The phosphate rock trade out of the West Indies island of Aruba had attracted many competitors, including sailing ships built in Brixham, Salcombe, Bideford, Barnstable, Padstow, Falmouth, Truro and Fowey. Full sailing ships were needed to carry freight over the 'western ocean' between Labrador, Newfoundland and Europe. The barquentine became the ship of choice. The bigger schooners, with their large, heavy, difficult to handle mainsail, started to be replaced and were converted to the lighter-rigged three-masted schooner class. The voyage from Cadiz to Newfoundland was more suited to the style of the new *C. E. Spooner*. During this lull in Porthmadog shipbuilding, ships were purchased from other ports and their builders. Vessels from West Country ports hoisted the Porthmadog flag.

The final phase of shipbuilding in Porthmadog started in 1893, and the records of the insurance society continue with precision. Many members of the society were then relatives – sons, uncles, nephews – of the original founders. Many were men who had served on and commanded the ships they were insuring. At this date, 100 vessels were insured locally. Mr Ebenezer Roberts, the experienced shipbuilder, was still on the committee of the society in 1899, embodying the wealth of experience the society enjoyed through the years, and its continuity of officers.

Despite a recession in the slate trade in the 1880s the trade picked up again at the end of that decade and in 1892 exports from Porthmadog reached their highest level – 98,959 tons by sea and 54,878 tons by rail. This revival saw new investment in shipping and two ship designers, David Williams and David Jones, emerged to build Porthmadog's ultimate type of sailing vessel – the Western Ocean Yacht.

14

Building the Finest Ships

Survey reports on the later vessels ... show that the Porthmadog vessels of the 1890s and the first decade of the present century were superior both structurally and in the materials of which they were constructed to their contemporaries and, indeed, to many of their predecessors. The later Porthmadog schooners were, indeed, as Aled Eames says, quite outstanding vessels, the ultimate development of the small wooden merchant sailing ship in Britain.

Basil Greenhill
National Maritime Museum, London

Apart from the lull in the 1880s – only the *Richard Greaves* was built in this decade – shipbuilding continued to exert its thrall on the coastline of Porthmadog. There was a resurgence in 1889 and the final stage of shipbuilding commenced. These ships incorporated all the design and build improvements that came from seventy years of local shipbuilding. The Western Ocean Yachts were inevitably of similar design, dimensions, building processes and materials. Mostly schooners, they incorporated the curving bulwark; the flattened stub at the stern with the sculpted surface below, the higher prow with a sharp, backward-sloping edge underneath; the sturdy bowsprit and strong, very tall, masts. They sat high in the water, except when they loaded with slate or heavy cargo: this is what kept them upright and stable. Unfortunately, their virtues were sometimes their vices. Because they were built for speed, with a small draught and a large spread of canvas, they were subject to instability in high winds and many were lost in hurricanes on the oceans.

Two men dominated Porthmadog shipbuilding in the last twenty years of its active life – David Williams and David Jones. Their building yards were next to one another, on the oddly named Rotten Tare, a spit of land located on the southern side of the harbour which had expanded through the century by the dumping of ballast by incoming vessels. David Jones had the advantage because his yard was adjacent to the harbour mouth so he was able to launch his newly built ships directly into the deep water and wide channel. David Williams's yard, however, was further inland and on higher ground, lying square to the harbour, so due to the narrowness of the channel he had to launch his ships sideways on. However, both were remarkably industrious and built new vessels yearly until 1913. In his splendid booklet *Ships and Seamen of Gwynedd*, Aled Eames lists the western ocean yachts built by these two stalwarts and two others. Three were completed in 1891 by David Jones with one built by Griff and David Williams. Ebenezer Roberts built the *Consul Kaestner* in 1892. In all, thirty-three of these magnificent vessels were built between 1891 and 1913, with David Williams building the final one in 1913, the *Gestiana*.

This vessel in construction is either the *R. J. Owens* or the *Inallt II*. David Williams, shipbuilder, is standing to the right wearing a topic helmet. One man is holding a hammer, another is holding an adze. Four of the team are wearing white hats, copying their boss.

We have the advantage of having good pictures of David Williams; a family portrait shows him with his wife and five children, three girls and two boys. He is wearing a modern-type white shirt and tie, an older-fashioned long waistcoat with lapel and a jacket with three buttons, no doubt made by a local tailor. His face displays a bushy moustache and mutton chop whiskers. He engages the camera with a straight look, steely eyes and a half-smile. He exudes confidence and success. He is also pictured at work during the building of the three-masted schooner, the *R. J. Owens* at Rotten Tare in 1907. He stands slightly aside from his workers, his mutton chop whiskers visible, wearing a hat of the trilby type. In another photograph, unusually of the inside of a ship being built, he stands wearing a 'topi' helmet, his eight workers with him, one carrying an adze, another a long-handle metal hammer, another a hand drill. It appears that some of his men sport wide-brim light-coloured hats (unusual at the time) in a copy of their boss's headgear.

Pictures of David Jones, however, are much harder to find. There are no pictures of him, as far as we know, in a photographer's studio, or of him posing with his family. There are pictures of him

A unique photograph, taken *c.* 1900, showing massive oak timbers. David Williams is to the far right. (Courtesy of Gwynedd Archives)

with his workers, which is where we feel he thought he belonged. There is one picture of him taken during the construction of the schooner *Gracie* in 1907. He stands in a line of ten men, a stout man wearing a bowler hat. In other pictures he is in the centre of activity, hardly distinguishable from his workers. He had a larger team of workers than his neighbour in trade David Williams. One gets the impression that David Jones was more of a hands-on man, less interested in drawing and planning and more inclined to a 'build it and see' procedure.

Building and launching ships was a social activity. It involved the whole community, as every street housed maritime workers and suppliers. Talk was of ships, owners, voyages, faraway places and sometimes, misfortunes.

Eyes were turned to the Rotten Tare/Cei Newydd shipbuilding yards of the two Davids, and when a ship was nearing completion those men who intended to sail on the ships had a sense of pride, looking forward to regular pay despite the discomforts of accommodation, mediocre food and the violence of the sea. Periods at sea would stretch from two or three months to as long as nine months, and sometimes longer. Some periods took the sailors around Europe, especially to Germany, Spain and Italy, a number of times. Other periods included very long trips such as all the way down the Americas, around the Horn, and up the coast of Chile (*c.* 2,800 miles) as far as San Francisco. It could be that sailors were away from home for two or three years.

There was something new and challenging about it all. The dangers were downplayed and the tone was one of excitement and optimism, something which was encouraged by reports in the local papers. In 1895, two newly built merchant ships, the *Sidney Smith* and the *Elizabeth Llewelyn*, built by David Williams and David Jones, were launched on the same day:

On Tuesday morning last two vessels were launched in the harbour amidst the cheers of a large crowd of spectators. It is not simply for the pretty sight of witnessing the vessels slide gracefully from the stocks to take their place so buoyantly and gladly in their new home that the great number of people attend these launches; but everyone feels that it is the surest token of the business prosperity of the place, and the news of the placing of a new vessel upon the stocks is always received throughout the town with pleasure.

The *Sidney Smith* voyaged to a large number of destinations. She went from Preston to Gibralter with coal; Cadiz to Newfoundland with salt; from Rio Grande du Sol to Falmouth with a load of raw hides (a stinking cargo); from Falmouth to St Petersburg with stores; from Russia to Perth, Scotland with a cargo of linseed cake; from Falmouth to Gibralter with a load of granite blocks. Richard Williams of Terrace Road, Porthmadog, served on her as an ordinary seaman and then as her captain: he told Emrys Hughes that his ship arrived in Gibralter with thirteen other ships following in the next 24 hours and that none had sighted one another in their passes across the Atlantic.

The *Sidney Smith*, a three-masted schooner, owned by eight men, mostly local, was lost in Twillgate harbour, Newfoundland, in December 1912. She had lasted seventeen years.

The *Elizabeth Llewelyn*, built by David Jones in 1904, was also a three-masted schooner. She was a total wreck on a passage from Gibralter to Hualva on 8 February 1912, after being stranded near Pearl Rock, Gibralter. Her certificate of registration was lost with the vessel.

The hazardous nature of the voyages of Porthmadog ships is clear, as they were often wrecked at sea. Many lives were lost, although there are accounts of all hands being saved by passing vessels or by the crew swimming to shore.

From a historical perspective, regrettably not a single Porthmadog ship – and over 250 were built – has survived intact. It was reported that one such ship lay in the harbour of Port Stanley in a poor condition, but was destroyed during the Falklands War. However, timber remains of the three-masted schooner the *M. A. James* (built at Porthmadog in 1900 by David Williams) lie on the muddy banks of the River Torridge in Devon. She was last reported on the River Clyde in 1944. At one point in her early life her major shareholdings were owned by four men of Borth-y-Gest. A photograph of this hulk appears in the *North Devon Archaeological Journal* of spring 2007 in an article by Chris Preece. Her huge oak timbers are visible above the mud as is her distinctive squared-off stern and the sculpting of her timbers underneath.

The remains of the *M. A. James* in the River Torridge, Devon. (Courtesy of Chris Preece)

15

Hiring Sailors: Voyages

As Porthmadog grew in importance as a port with newly built sailing ships, so the need for manpower grew. Hiring sailors for work on merchant ships became commonplace, and these hands came from the immediate area but were also recruited from a wider area, including Caernarfon, Bangor, the Lleyn Peninsula, Harlech and Bala. Typically, young lads came in search of work from the rural areas, where paid work was scarce. The quarries of Blaenau Ffestiniog sucked in labour, as did the slate work in transporting, stacking and loading slates on the new quays at Porthmadog.

A typical Porthmadog sailing ship carried seven hands: the captain, mate, cook, two able seamen and two ordinary seamen. The youngest was sometimes a young man of fifteen or sixteen. It is extraordinary to think that these vessels traversed such long distances with only a few sailors on board, given that the rig was a complex system of canvas, ropes and pulleys, which had to be altered frequently with changing wind and weather conditions. In the southern latitudes, and sometimes in the northern latitudes (with cargo to St Petersburg), canvas and ropes would become frozen and difficult to handle.

A seaman was hired for a voyage lasting a number of months and destinations could change en route depending on where there was cargo to be collected and delivered. A twelve-month voyage was not unusual, and some voyages lasted two years or more. To sail from Porthmadog to Hamburg, Germany, in about five days was usual. Additionally, Mediterranean ports were often called in, especially Cadiz, Naples and Genoa. North Africa was also popular, especially Morocco. The 'western ocean' was regularly crossed, the east to west voyage typically carrying salt, with the reverse journey carrying self-packed cod for the Catholic countries bordering the Mediterranean.

Slate was carried in many directions across the world. The European mainland was accessed via the River Elbe. In addition to Europe, many ports on the eastern seaboard of the North American and South American continents were visited by Porthmadog cargo ships, for example Rio de Janeiro was visited. Some voyages included rounding Cape Horn, braving its violent weather conditions, sailing the vast coast of Chile, with San Francisco as the final destination. This mammoth voyage saw sailors spending months at sea, sometimes seeing no other vessel, covering over 12,000 miles from home to destination. These, of course, were the days before the Panama Canal.

The voyages of the *Elizabeth* can serve as an example of a Porthmadog's ship's merchant life at sea. She was known, according to Henry Hughes (p. 195), for having sailed across the Atlantic, west to east, from South America to Liverpool in twenty-three days. Her rival, the *Blodwen*, sailed from Labrador to Greece in twenty-two days. In 1895, under Captain Evan Jones of Porthmadog, the *Elizabeth* made the following voyage:

- Porthmadog to Stetin, in the Baltic, with a cargo of slates
- Stetin to Rochfort with oak planks for the manufacture of coffins
- Rochfort to Cadiz in ballast
- Cadiz to Harbour Grace, Newfoundland, with salt
- Labrador to Gibraltar with fish
- Leghorn to Retino, Crete, with hundreds of tons of fresh water
- Retino to Goole, England, with olive oil

The *C. E. Spooner* sailed from Shoal Bay, Labrador, to Liverpool, passing Fastnet Rock, Ireland, in thirteen days. She did the journey in reverse in eighteen days. She was on an eleven-month voyage, the details of which are:

- Porthmadog with slates to Hamburg
- Hamburg to Cadiz
- Cadiz with salt to Newfoundland
- Harbour Grace to Shoal Bay, Labrador
- Shoal Bay with fish to Liverpool
- Liverpool to Garston
- Garston with coal to Gibraltar
- Gibraltar with ballast to Huelva, Spain
- Huelva with copper ore, to Exmouth
- Exmouth, with limestone, to Porthmadog

These voyages illustrate the long time from home that the mariners had to endure, and the varied, and often very unpleasant, conditions which applied at sea. Also, the cargo carried was varied and difficult to bring on ship and to store. Mariners had to be strong and resolute men.

Porthmadog-built ships were built to last. Their classification A1 lasted for the ship's first twelve years of life. A major overhaul was a time-consuming and expensive procedure; rigging had to be replaced and the ship's timbers were removed at intervals so that the condition of the hull could be examined. Deep-water vessels carried a skin of copper below the waterline, with felt underneath, which would be replaced when not in A1 condition. Hundreds of copper nails had to be removed and replaced. It was industrial work and occupied many hands in Porthmadog's port for many weeks' work on each ship. This 'provided full-time employment to an army of carpenters, riggers, block-makers, ship smiths, joiners, sailmakers and copper workers, not to mention the ship chandlers who re-fitted the vessels with ropes, hawsers, and the thousand and one items of ship's tackle.' (Henry Hughes, *Immortal Sails*, p. 176).

Henry Hughes has his recollections of the Porthmadog shipbuilder par excellence David Jones:

David Jones, always a busy person, had a galvanic nature – one of those men who, should he have a ship to build, seemed ever on the spot and at his work. Watching him from our garden, which enjoyed a birds' eye view of the famous yard, his familiar figure, with his bowler hat placed on the back of his head and plans bulging out of his pockets, could be recognized dashing about from one gang of workmen to another, leaving nothing to chance. Enjoying robust health and an overwhelming share of vigour and vitality, he had a way with him which brought his men to the same pitch of enthusiasm and sense of duty as he himself possessed. Although the new ship, which took the name of *Richard Greaves* was a little gem, except for the excitement of the launching ceremony, the occasion failed to stir the enthusiasm of the people sufficiently to influence them to embark on further building commitments. But there was not long to wait. Urged by the rapidly-increasing and ever-widening gaps in the ranks of the deep-water ships – the eighties took their heavy toll – Porthmadog awoke from its twighlight slumber like a giant refreshed. The autumn sun of its life played encouraging rays of light on the old building slips again. A rebirth of confidence brightened the faces of the maritime minded. Soon the giant hulls of little ships towered above the quays once more, and there was much animation in the port; the year 1889 opened a new chapter of prosperity in its life. (*Immortal Sails*, p. 176)

The most common voyage made by a Porthmadog-built ship was from home to Hamburg with slate in five days. But the record was held by the *M. A. James*, sailing the 1,000 or so miles distance in four and a half days.

c. 1900. Fellow Porthmadog mariners have a drink together in a remote spot in the north of Chile. Captain Richard Griffiths is one of them. This location is Caleta Coloso. The coastline of Chile is approximately 2,800 miles long; they had endured Cape Horn in sailing ships with no engines and with usually only six men onboard. The trip from Porthmadog to San Francisco was at least 12,000 miles.

The other common trip was for phosphate rock to the island of Aruba in the West Indies. The *W. W. Lloyd* (built in Porthmadog in 1875 by Ebenezer Roberts) did it in twenty-seven days, but the three-masted schooner *Elizabeth* in 1906 beat that with her twenty-three days.

Perhaps the most remarkable trip of all made by a Porthmadog ship was by the *Blodwen* in 1901. She started a race with a Dumfries clipper, from Tickle, Labrador to Patras, Greece, a distance of 4,500 miles. On the twenty-second day, this schooner sat in Patras harbour, having made the run of over 200 miles a day in twenty-two consecutive days.

Perhaps the longest voyage by a Porthmadog vessel, the *C. E. Spooner*, (a three-masted schooner) took two and a half years and covered some 29,000 miles. This included a voyage from Cadiz to South America with salt, a journey of 4,500 miles; a return jouney from Rio Grande do Sul to Falmouth with raw hides (the smell!), then on to St Petersburg with hides, covering 2,000 miles. There are three entries on this ship in Fox's Arrival Books – they were principal shipping agents in Falmouth. One was in September 1901, from Rio Grande to Havre; one in April 1903, from Rio Grande to Havre and the third in August 1911, from Rio Grande to Altona (River Elbe, Hamburg). The cargo on each occasion was hides and horns. The men onboard were her master, mate, two able-bodied seamen, one ordinary seaman and the cook – only six men in total. Their pay, in the years of the turn of the century, was £7 or £8 a month for the master; £4 or £5 for the mate; the ABs £3 10s; the OSs fewer than £3.

16

The End of Sail

As previously said, the ship *Gestiana* was built as the last of the western ocean yachts in 1913. She was built by David Williams on the centre part of Rotton Tare, following the final ship built there by David Jones, the *Elizabeth Pritchard*. She was wrecked in a fierce gale off Newfoundland in October 1913 and her certificate of registration was lost with the vessel. There had been a tragic air about her as the shareholders had gathered to see her off and the christening bottle failed to break. In the previous year, many Porthmadog ships had come to grief, and after her launch, more ships went down, including the *Blanche Currie*, which was last seen on 4 February 1914.

The Portmadoc Mutual Ship Insurance Society was in trouble. Many ships had been lost, incomings were slowing and outgoings were mounting. Exports to Germany ceased. Steam power began to take over the cargo trade. As Henry Hughes writes:

> The end of a romantic and glorious era had come. With the Portmadoc Mutual Ship Insurance Society closing its doors in the summer of 1917 the heart of maritime Portmadoc ceased to beat. It had pulsed with vigour and strength for close on a hundred wonderful years. (*Immortal Sails*, p. 240.)

17

The Festiniog Railway Before Steam

The Royal Assent for the Act which enabled the new line from Blaenau Ffestiniog to Porthmadog to be built was awarded on the 23 May 1832. It was to be 1 foot, 11½ inches gauge.

The first stone of the railway was laid by W. G. Oakeley of Tanybwlch on the 20 April 1836. When the line was ready for use a train of wagons laden with slate from the Samuel Holland quarries came down the line. By this time Henry Archer had created the Festiniog Railway Company. The quarries at Rhiwbryfdir had not agreed to have their slates carried down by rail so Mr Holland was the only owner to use the rail until 1839, when the Welsh Slate Company also began to use the rail.

The capital of the Festiniog Railway Company was £24,185, although it had borrowed £14,000 to enable it to complete the railway construction. It was profitable, paying a dividend of 6, 7 and 8

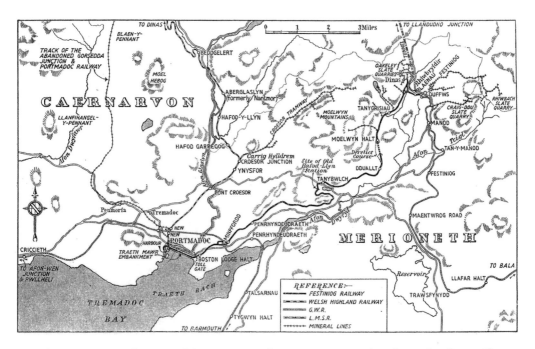

An early map showing the route of the Festiniog railway, starting at Porthmadog and ending at Blaenau Ffestiniog. It is approximately 13 miles long.

per cent a year. It grew to conveying from 45,000 tons of slate to 50,000 tons of slates annually to Porthmadog in the early 1850s. New quays were built by the Welsh Slate Company, Rhiwbryfdir Slate Company and Mr J. W. Greaves. In October 1838 William Madocks's daughter, then Mrs Roche, paid a visit to the port and was greeted with enthusiasm, including a public reception.

We look back now at some of the salient detail of the building of the track from Porthmadog to Blaenau Ffestiniog and we view the matter with some puzzlement and considerable respect. It is accompanied by such drama that one could make a successful film out of its travails. It was a huge undertaking, and the fact that it was completed in a little over three years seems almost incredible.

The contract of 22 December 1832 begins, 'The Railway to be formed for the reception of Iron Rail, according to the plan or Specification furnished by Mr Spooner.' This original contract for the work made such demands on the contractor that it is surprising that any contractor would take it up and agree to its terms. The work expected was extremely arduous and time-consuming, much of it consisting of cutting by hand through rock and stony ground.

The original contractor, James Smith of Caernarfon, started by agreeing to be paid £6,972 for the work, to be paid in instalments against every eighth of the work completed. The Festiniog Railway Company would also pay the engineer and foreman. The length of the railway was expressed at 14 miles (in fact, 13½ miles is closer to the truth) and the contractor was required to purchase the necessary land. The contractor was also to acquire and lay down iron rails and chairs, and use proper stone blocks and sleepers as base. 100 wagons each to contain 23 hundredweight of slates were also to be supplied by the contractor for the 'use and benefit of the said Company'.

Not surprisingly, by April 1834 Smith had given up on the project, but in two years he had managed to get most of it done, but not to a standard insisted on by Archer, who dismissed him as contractor. The dispute went to court (not an unusual occurrence in the days of railway investment and building), with Smith alleging to have completed seven eighths of the work and the company saying that he had only completed three quarters. The arbitrators awarded nominal damages of 14s 4d against the company.

The company stepped in and determined to finish the line themselves, raising fresh capital of over £24,000. Alleged shoddy workmanship created by Smith cost over £4,000 to put right and another £5,209 was spent on necessary work. Pay sheets from 1834 show that James Spooner, as the engineer, was paid £100 per year. Thomas Pritchard, the foreman, was paid £65 per year. Over 150 men of different trades were employed; William Williams (senior), the father of the future William Williams who became superintendent of railways, joined the railway in the early 1930s and almost certainly was part of this mammoth groundworks exercise. Pay sheets reveal the town of origin of the workers: they were drawn from a wide area (including Caernarfon, Holyhead, Pwllheli, Bala) and comprised up to nine gangs of bricklayers and labourers, each gang having in the region of ten men. They were paid per piece of work; they dug cuttings and drains, building walls and embankments. It was heavy physical work, with picks, shovels, wheelbarrows, horses and carts. Local people sometimes were paid by the day, and workers from a distance found lodgings locally.

The original fish-bellied rails came from Dowlais and from Jevons Sons & Co. of Liverpool. The chairs (rail supports) were cast by Thomas Jones of Caernarfon who set up a new foundry at Porthmadog, later to be the Glaslyn Foundry. The slate and granite sleeper blocks were cut out of nearby quarries.

In 1836, for the period to 30 June, the first tonnage of slate was carried by the new railway when 1,195 tons of slate was carried across the Cob at a payment to the Madocks estate of half a penny per ton. Donkeys and horses provided power. The company paid Robert Stevenson £21 in 1835 for a design of inclines over the top of a spur at the Moelwyn mountain. These were put out of use in

May 1842 when a tunnel 730 yards long, started in December 1839, came into use. A celebration took place on 24 May 1842 to mark the completion of the work. The building of this tunnel, all done by hard labour with picks and wheelbarrows, is a story in itself.

By December 1834 money was short, but by July most of the trackwork was completed. However, building a lower road across the Cob was not. The following years saw friction among the board of the company, with Henry Archer at odds with others. However, he continued as a director of the company, being granted an annuity of £100 a year in 1860. He is known among philatelists as the man who invented a machine for perforating sheets of postage stamps. He died in south-western France in March 1863.

Looking back over the building of the railway, it is clear that James Spooner, Thomas Pritchard and Henry Archer did most of the managing work, with Thomas Pritchard the hands-on man, on-site daily, organising gangs of men and directing operations.

By June 1836, the lower road over the Cob had been completed. The opening ceremony occurred on 20 April 1836. A 'well-selected' band from Caernarfon played, rock cannon were set off and discharged 'from the fort at Morfa Lodge, the residence of James Spooner, Esq.' The railway's workmen were marched behind the band to Morfa Lodge where they were treated to a 'substantial repast on the lawn at the expense of the railway company' (p. 25, *Narrow-Gauge Railways in North Wales* by Charles E. Lee). The spirit is captured by Samuel Holland: 'At Port Madoc we were met by crowds of People, Bands playing and the workmen had a good dinner given them; all the better Company were entertained at Morfa Lodge by Mr Spooner and ended the day with a Dance.' Was James Pritchard invited to this dance?

'Better Company'!

Were not the labourers, masons and so on, who built the 14-mile twisting and rising railway by hand through rock and upland terrain in three and a quarter years also 'Better Company'? We look at this through the distorting prism of history.

In August 1834 Samuel Holland was the first quarry owner to use the new Festiniog railway. In the meantime a squabble had occurred over the building of railway-side stone walls and their height. W. G. Oakeley had insisted that for 500 yards, where the railway passed by his land near his house, walls should be built 10 feet high, and for the remainder of the distance on his land, walls should by 4 foot 6 inches high. Holland protested, writing to Oakeley: 'If the stone walls are built throughout, the public would be deprived of one of the finest Panorama views in North Wales.' Holland told Oakeley that the profits from the railway in carrying slate for his company and the Welsh Slate Company would well compensate him, and offered Oakeley £500 in compensation for the railway intruding into his view. However, Oakeley died in October 1835. His widow dropped the wall-building provision and accepted £500 from the company. Welsh Slate Company slates were carried in October 1838, followed by Lord Newborough's. Turner and Casson did not use the railway until 1860. Even though slate continued to be carried down the Dwyryd, the railway increased the amount of slate quarried, from 11,396 tons in 1825 to 89,294 tons in 1865. The railway soon declared a profit and paid its first dividend against shares, 7 per cent, in 1843.

The railway made heavy use of horses. They pulled the rolling stock up the line of some 13 miles, against a drop of 700 feet. Coming down the line, the horses were carried in a dandy at the rear, and the gradient (1 in 80 most of the way) was such that it could roll down all the way from Blaenau Ffestiniog to Morfa Lodge; then it would be pulled by horses across the Cob. It was no life for a horse. Coming down the line, brakes had to be applied and this was done by two men sitting on top of the slate wagons, sometimes leaping from one to another; a dangerous job. However, the heavy wagons were often reluctant to start their journey downhill, and men called pushers put their weight to the wagons to get them started.

In 1838 the cable incline was abolished and a decision made to drive a tunnel, as originally planned. It was 730 yards long, only 8 feet wide and 9 feet and 6 inches high. In July 1838 the company raised new capital of £12,000, with £4,000 in loans.

Thomas Pritchard worked for the Festiniog Railway Company in a managerial role, and he witnessed in his own hand an agreement created by Spooner with Morgan Jones of Rhiw Goch Farm dated the 6 November 1838 to provide horses and wagons.

The journey time for the length of the railway was about 2 hours down, but the fastest time was set up in the 1850s with 1 hour 32 minutes, from Quarry Terminus to Boston Lodge, with three stops. Going from Boston Lodge to Quarry Terminus took about 6 hours, each train being in four sections, each hauled by a horse and made of eight empty slate wagons, plus the dandy. Speed on the down train was restricted to 10 mph. A down train could be over 300 yards long, comprising over one hundred wagons. A travel book by J. Hemingway of 1835 includes,

> a most delightful ride, winding round mountains through deep cuttings in the rock, and thickly planted wood, and over precipitous vallies [sic], fearful to contemplate in the transit ... [Visitors] have the opportunity of enjoying this high treat in perfection, and without personal fatigue, as a carriage has been placed on the line connected with the Oakeley Arms Hotel. The quarries, with which the railway is in connection, produce the best and most valuable kind of slate, which, in consequence of its freeness from spots, may be termed *virgin slate.*

Meanwhile, at the source of this valuable material people lived simple, frugal lives in the high level of rain that descends on the Ffestiniogs. They had wages of between 12 to 16 shillings a week. Families had many children, often ten or twelve. Illness among quarrymen was common and some workers were crippled for life, sometimes due to accidents at work in the absence of health and safety provisions and typhoid broke out in the '30s and '40s.

However, this life was under-ridden by a self-regard, integrity and pride. Men worked for their families, their community and for their country, which was Great Britain, of which Wales was a part. Their sense of Welshness was anchored in the language and in intellectual culture. They valued skill with language and discussion and debate were paramount. In the next century, a disproportionate number of professional teachers came out of the Welsh quarry and mining towns, many heading for the main towns and cities of England, carrying with them the value their ancestors had placed on learning, literacy, expression and debate, much of it political. The institution of the Caban was part of quarry life; this was a name for a group of workers who would gather in their breaks from work to create discussions about affairs of the day. Minutes of these meetings have survived, showing how formally set up they were, with debate motions, a secretary and treasurer.

Religion played a strong part in quarry life. Nonconformist chapels of differing sects were built by hand, many in the artisan communities. Their language was entirely Welsh, with the Bishop Morgan Bible at the centre. Sunday was reserved for religious observance; a preaching event in the mornings and evenings and a quasi-school event in the afternoons, with children and adults attending Bible study groups.

This society of workers, the society which included the Caban, the chapel and education, all using the Welsh language, was separate from the management of businesses. Very rarely did a Welsh person become a general manager or company director, as they were not in possession of capital, did not inherit valuable property or were not in a position to raise it from bankers or shareholders. Thomas Pritchard and William Williams rose to positions of authority but were never made company directors. Board members were English-speaking, from the gentry of England, and usually from wealthy families, although some were Irish, like Henry Archer. Indigenous workers who started early in the quarry or railway rose, through their natural ability, to the position of manager or section head, such as William Williams, an engineer, who became superintendent of railways at the Festiniog Railway Company in 1886 until 1906, living in Boston Lodge and controlling all of the complex work of making locomotives, rolling stock and keeping the railway workings in good order. He was Welsh speaking, a writer of poetry, a lover of choral music, and unusually, attended the Anglican church in Penrhyndeudraeth. Without him and his skilled team of engineers, the essential improvements in the design of other-built locomotives, and in the construction of wagons and carriages, would never have happened. The improvements in the design of the Fairlie engines were done in Boston Lodge. In other words, it was the workers who made the Festiniog railway a success.

However, the well-being of workers was not entirely neglected and there are many examples of enlightened philanthropy. Samuel Holland Junior and John Whitehead Greaves were both Liberals in politics, not Tories. Holland was Merioneth's second Liberal MP and William Turner started an elementary school for quarry workers in Blaenau Ffestiniog. In 1864 a silver band was started by the owners, which became famous as the Oakeley Silver Band. Dr Willliam Homfray was established as Blaenau's first health officer in 1848. Quarrymen and their families had a very limited diet, much of it bread, butter and tea. John W. Greaves' daughter-in-law, Marianne, John E. Greaves's wife, fought a twenty-year battle to improve the eating habits among the workers, urging them to eat vegetables and fruit as well as meat for protein. However, a tight domestic economy prevailed.

The engineering gang at the Festiniog railway manufacturing shed with handmade *Topsy* on display. William Williams, foreman, is to the right in the back row.

18

Steam Engine Power

The Festiniog Railway Company continued with the idea of horsepower. In 1850 Charles Easton Spooner, son of James Spooner, reported to the board that 6 miles of track had been relaid with heavier metal. The coming of locomotives was in the air, not only to carry slate but also workers and visitors. On the death of his father James in August 1856, Charles became 'future Manager and Clerk'.

However, putting steam on a narrow-gauge rail had never been done before, and many said it was not possible, especially on such a line as existed at the Festiniog railway. The central problem was corners: how to pull a long train with a steam engine around close curves. The other problem was power: how to create enough energy in the locomotive sufficient for it to pull heavy wagons up a long 1 in 20 gradient. Brunel was against it and Robert Stephenson said it was impossible. However, at a board meeting in August 1860 Charles Spooner was delegated with the task of researching and making recommendations on the matter of adopting steam power. Neath Abbey Ironworks, south Wales, had working narrow-gauge locomotives which Spooner went to see, reporting his findings in November 1861. Holland had argued that they should procure such an engine, and if necessary 'rectify defects, etc.', which shows how far the engineering section of the railway had progressed.

On 10 October 1862 the periodical *The Engineer* carried an advertisement from the Festiniog Railway Company asking for companies to contact them with a view to manufacturing locomotives. They had twenty-nine responses. However, none were given the contract, which was granted to George England and Company of London. Charles Menzies Holland had influence in this decision. In April 1862 the board specified that two engines were to be ordered but in March 1863 they ordered three – two to be delivered by 1 June 1863 and a third to be delivered before 1 March the following year. These were tank engines, four-coupled, with tenders for coal carrying.

In 1863 two new engines had been delivered by 28 July. The first was carried to Porthmadog from Caernarfon by Harry and Job Williams of Minffordd, drawn by four horses, on 18 July. It was lifted and placed on the railway line and pushed by hand across the Cob to Boston Lodge. The first two locomotives arrived in July 1863; the third early in 1864 and the fourth in March.

Boyd writes:

It is known that two engines had been delivered by 28 July 1863. These are now thought to have been *Mountaineer* and *The Princess*, carrying their original numbers No1 and No2 respectively ... In March 1863 the Company had agreed with England that two engines should be supplied at £1,000 each and £800 for a third, the last to be held back in case modification should be necessary. (*The Festiniog Railway*, Vol. 1, p. 33)

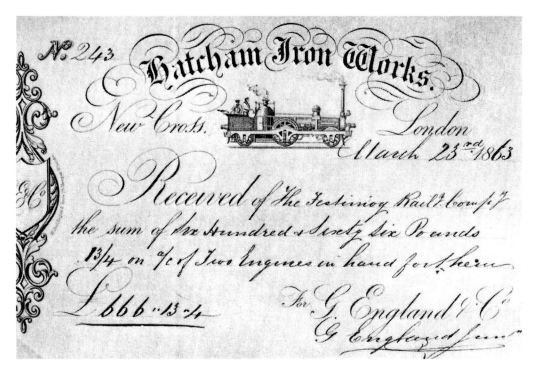

23 March 1863. The Festiniog Railway Company paid £666 13s 4d for two engines to be manufactured by George England of the Hatcham Iron Works, New Cross, London. This was a part-payment.

The third engine was the *Lord Palmerston*.

The first trials of the new locomotives did not go well. *The Princess* was sent on a short trip in August 1863. Looking at the faults, Holland admitted, 'I own it is a rather ticklish thing to manage.' However, the engineering shed at Boston Lodge had taken a good look at the engines and modified them under their own hands. The *Railway News* reported, 'The engines, which only weigh 5 tons, are the smallest ever made for railway traffic. They are beautiful ... tank engines.'

By September 1863, after fitting both engines with a larger dome, and other modifications by the Boston Lodge engineers, the engines were fit to travel and Charles Spooner penned the following invitation to a selected guest list:

Bron-y-Garth, Portmadoc
October 16th, 1863

The Festiniog Railway Company propose opening their Line for traffic by Steam on Friday 23rd instant, and intend starting a Train from Port Madoc at 10, on the morning of that day to the Quarries.

They hope to have the pleasure of your company up the Line, and return to Dinner at 3 o'clock at the Town Hall, Port Madoc.

Richard Richards of Bangor wrote an eyewitness account:

Two trains started from the Portmadoc station, with an engine attached to each, the number of persons altogether being about two hundred.

The reader need not be possessed with any lively fancy to imagine to himself the laughable incidents which would be likely to occur on a railway line when twenty carriages or so, for the first time since the creation, bounded along faster than any stage coach and with no visible propelling power; for there were hundreds of people in the neighbourhood who had never seen a steam engine although of course they had heard of one. Horses galloped about the fields like distraught animals, when the puffing engines passed along, so confused and amazed were they at the noise, and the phenomenon altogether and the more timid cows and sheep seemed equally as astonished. All along the rails, at every available spot, crowds of wondering people were collected in groups, from the aged crone of ninety years, to the demure little damsel of three. When the trains arrived at the terminus at Blaenau Ffestiniog, we were greeted by hundreds of quarrymen, who were perched on the rocks many yards above us, who cheered lustily and uproariously, and as only Britons can. The engines were piloted by Mr Holland and Mr England, and they must have been deeply gratified with the successful results of their skill and labour. (*The Little Wonder*, p. 54)

In October 1864 a Board of Trade inspector, H. W. Tyler, made a number of recommendations, which were carried out, and on 5 January 1865, the company had an official opening. Passengers, officials and guests were carried to Blaenau Ffestiniog. It was marred when an engine cleaner jumped off the train at Boston Lodge and was struck by the first carriage: he later died. Classes of passengers were created and uniforms supplied to the staff. Reverend Tim Phillips wrote in the *South Caernarfon Leader* of 1946 about stationmasters' attire:

This important official was arrayed in a vesture which consisted of a frock coat, waistcoat and trousers, of dark navy-blue cloth, excellent in quality, and a cap of the 'cheese-cutter' pattern – all adorned with thick gold braid. He walked up and down the platform with grave dignity, carrying a baton (the staff) in one hand and flourishing a carriage-key in the other. Not less spectacular was the uniform of the Passenger-Guard. Early photographs depict him as wearing a well-tailored frock coat. The cap was of a different pattern – one similar to the one worn by the Czar of Russia. His uniform was adorned with silver braid, and he carried a broad, highly-polished, morocco strap across his right shoulder, which formed the receptacle of a big silver chronometer. (*The Little Wonder*, p. 56)

The third and fourth engines were, according to John Winton (p. 53) *Mountaineer* and *Palmerston*. He asserts that *The Prince* was No. 1. However, Boyd has *Mountaineer* as No. 1. The original engines were found to have insufficient adhesion, so the design was changed and enlarged in 1867, and George England & Company produced No. 5 *Welsh Pony* and No. 6 *Little Giant*.

The first series of four engines were too small and too lacking in power for the weight of traffic that applied. A fourth and fifth engine were ordered, larger and more powerful than their predecessors: they were the No. 5 *Welsh Pony* and No. 6 *Little Giant*. Larger driving wheels, extended footplates and cylindrical sandpots, designed and made by local blacksmith John Williams (Ioan Madog), William Williams's uncle, were featured on these locomotives. The wheelbase was longer than on previous models, and the boilers 6 inches longer. The driver of the vehicle was on the left-hand side, and the vital single-line staff was held in a special pouch on the driver's side.

Princess," the first engine of the Festiniog Railway as originally built

" The Prince," after being fitted with a saddle tank

The *" Little Giant,"* built by George England & Co. in 1867

EARLY FESTINIOG RAILWAY LOCOMOTIVES

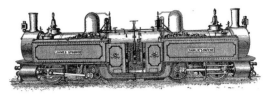

Double Fairlie locomotive " James Spooner " as first built. Note the bells on the sandboxes

The *" James Spooner "* as rebuilt with an all-over cab. Standard size tanks, sandboxes, and cast-iron chimneys have been fitted

Single-boiler Fairlie type locomotive " Taliesin," built by the Vulcan Foundry in 1876. The all-over cab was added later

FESTINIOG RAILWAY FAIRLIE LOCOMOTIVES

Above: Six illustrations of the Festiniog railway's early steam engines. Notice the difference between the *Prince* and the *Princess* and the later double-Fairlie *James Spooner.*

Right: The cylindrical sandpot designed by John Williams, blacksmith, of Porthmadog (poetic name Joan Madog) is still in use. Here it shows in detail as part of the *Merddin Emrys.*

The 0-4-0 type of steam locomotive enabled longer slate trains to operate and in turn this facilitated the introduction of passenger trains in 1865, making the Festiniog railway the first narrow-gauge railway in Britain to carry passengers. Following this, in 1869, the Festiniog railway's first Fairlie articulated locomotive was introduced and these locomotives, with their double ends, have now become one of the railway's most easily recognised features.

19

Robert Fairlie's Patent

In 1868 the management of the Festiniog railway, under C. E. Spooner, were convinced that the capacity of their railway should be increased, as revenues and traffic had increased substantially. In 1869 engine mileage was 50,314, and receipts for the year totalled £23,377, with total expenses being £13,054.

Robert Francis Fairlie was a Scotsman, born 1831. He developed a close interest in railways and engine design and was a superintendent of the Irish Londonderry–Coleraine railway when he was only twenty-two years old. He showed a close interest in the manufacturing activities of George England and married his daughter in 1862. His paper *Railway Management* was read to the Royal Society of Arts and then printed in the periodical *Engineering* in 1868.

Charles Spooner had taken an interest in the work of Robert Francis Fairlie and his double-bogie articulated locomotive since *Progress* was first introduced on the standard-gauge Neath and Brecon Railway in 1865. He saw that doubling the present Festiniog line to enable more traffic to be carried would be extremely expensive. The present traction system relied on a tender to carry fuel and water, and this had no driving wheel power. Also, because the present locomotives had a distinct front and rear, they required a turntable or large loop to turn them around for the return journey. The Fairlie locomotive (which looked like two locomotives fastened together) applied more power to the rails and thus could pull longer trains, and could run in both directions on the same rail. The Fairlie design carried fuel and water inside both engines and had each axle driven by steam power. There were two boilers in the locomotive, joined at the firebox end. Originally, as designed by Fairlie, the boilers shared a common firebox, but with separate water spaces. The Boston Lodge engineering team altered this and created two separate fireboxes. There were also two sets of driving wheels and cylinders set on a separate bogie (swivelling platform) that could negotiate corners and these were separate from the main engine frame. The driver worked on one side and the engineer on the other; the joined fireboxes separated them. There were controls on both sides of the cab to enable the engine to be run in both directions.

Fairlie was approached and agreed to design such a locomotive. It would be the first time his patent would be applied to a narrow-gauge rail pulling trains over a long gradient. It was built again in the George England works, the company now named the Fairlie Steam Engine and Carriage Company, and was named No. 7 *Little Wonder*. She was a 0-4-4-0 double engine. She had no cab, but weatherboards were fitted. Small ploughs were fitted at the front of the cylinders for clearing snow off the line. The words 'Fairlie's Patent' were inscribed at the side of the rear unit, and 'Little Wonder' against the front panel.

The cost of this engine was £2,006 but it recovered its cost through its fuel economy as it ran on 75 per cent of the fuel required for the earlier engines. This engine lasted some twenty years; when its boilers had worn out, it was withdrawn from service.

This revolutionary engine (a 'Double Fairlie') was delivered to Boston Lodge during July 1869. She was ready for trials on 18 September 1869, before a Board of Trade inspector. The *Little Wonder* had to draw 111 slate wagons as well as passenger coaches and goods wagons. She pulled this load without trouble up to Hafod-y-Llyn. The Moelwyn tunnel was traversed in 1 minute 5 seconds at an average speed of 23 mph. On the return journey, with a maximum speed of 35 mph, *The Engineer* reported 'the bogie engine swinging round the curves with graceful ease and a total absence of strain or jerk.'

Other trials were run in February 1870, and attended by the Imperial Russian Commission, which had been detailed by Emperor Alexander II to report on narrow-gauge railways in Britain. The adoption of the 2-foot gauge in different parts of the world (notably on the Darjeeling–Himalayan Railway) is largely due to this successful event. One celebratory event featured the miniature hand-built steam train *Topsy* (the size of a small terrier dog) which ran on a 3⅛-inch gauge at Bron-y-Garth, Spooner's house overlooking the estuary on the Porthmadog side of Borth-y-Gest. This was an extraordinary engineering feat by William Williams, who built the engine himself, using handmade tools, one of which, a spirit level in brass some 4 inches long, is held in the family of William Williams today. Parts of the rails still exist and are stored by the Festiniog Railway.

Fairlie was very keen to establish his name and to create locomotives of the new sort. He travelled abroad, including to Venezuela, where he caught malaria, to establish agencies. He was so pleased with the success of *Little Wonder* and the publicity gained that he gave the Festiniog railway the right to manufacture and use his designed locomotives free of patent charges. He died in 1885.

The front Fairlie, *Little Wonder* (No. 7) at Porthmadog station. Notice the rigging of a ship in the harbour to the left.

Little Wonder hauling a huge load at the curve at Creua as an Up train, near Tan-y-Bwlch. Mineral wagons are immediately behind the locomotive, with passenger coaches behind the wagons.

Various companies used his double-Fairlie design but the design and performance of *Little Wonder* was never surpassed.

In 1872 the Festiniog railway ordered another Fairlie locomotive to be manufactured by the Avonside Engine Company, the No. 8 *James Spooner*, designed by G. P. Spooner (Percy), Charles Spooner's son. It was larger than *Little Wonder* and became the prototype of the double Fairlies which the Festiniog Railway Company later built.

It was followed in 1876 by a tank engine locomotive, the single Fairlie No. 9 *Taliesin*, which had a 0-4-4 wheel arrangement and operated until 1927. By now Boston Lodge had a fully equipped railway manufacturing workshop and they built the new engines in-house: the No. 10 *Merddin Emrys*, built in 1879, and *Livingston Thompson*, built in 1885.

In 1869 Charles Spooner realised that the rails then laid were too light for the demands of the new trains so he had them replaced with longer, heavier, bullhead rails. He continued to publicise the virtues of narrow gauge, publishing a book entitled *Narrow Gauge Railways* in 1871.

Scribner's Man

In December 1878 a journalist called William H. Bishop, working for *Scribner's Monthly*, visited the railway and wrote a vivid account. He described James Spooner as 'a hale, dignified gentleman of sixty. His son, a young man of energetic and companionable traits, my guide'. Bishop took the early morning quarrymen's train to Blaenau Ffestiniog, leaving at 6 a.m.:

It was naturally much before daylight, at this hour of a mid-winter morning. Snow-flakes fell thickly at intervals. The quarrymen came trudging out of the silent streets with their ration of supplies in canvas bags knotted across their shoulders. They stamped the snow off their heavy boots in the station, and talked softly together in their strange tongue. Among them – the one touch of brighter sentiment in the scene – a rugged man, stiff in the joints from toil, had beside him a pretty child, a girl of ten, who carried in a satchel a part of his provisions. She was shabbily dressed, as became a quarryman's daughter; the small face was rosy with the storm, and the unkempt blonde hair had a genuine interest even apart from her circumstances. His only reply to compliments was the common *dim Saesnach* [sic] (no English) impassively spoken ... The engine at the head of the long train of red-painted boxes slipped on the icy tracks and did not easily get under way. While it fumed and shrieked in the rage of ineffectual efforts, telegrams came down the line countermanding the train ... [Little Wonder had] an odd aggressive-looking build ... driven by one Williams, head of the machine-shops and the most trusted mechanic on the line ... The small wooden station at the end of Madocks's embankment has hardly more than the look of a sentry-box ... [tickets are of] the usual size, and porters, guards, and brakemen bustle about with an important air as if they had never been out of the service of Isambard Kingdom Brunel.

Bishop goes on to describe Blaenau Ffestiniog:

The quarries are vast abysses, gloomy as the pictures in Dante's Inferno. Slate is everywhere. It strews the slopes with debris, turning the light with slight blueish reflections; is set in to the tops of walls instead of broken glass; and stands in irregular slabs like tomb-stones, for fencing around outlying houses. The homes of the quarrymen are for the most part in barrack-like structures on the heights. They are little given to revelry, and there is little in the villages to attract them down if they were. We passed the yellow van of a travelling show at one point, lying deserted in a field, melancholy as a grass-hopper in winter. Both horses and proprietor had turned out to work in the quarries.

Bishop wrote of the challenges to the Festiniog railway by the standard-gauge trains. These were the North Western and the Great Western. The London and North Western reached Blaenau by a 2-mile tunnel under the Crimea. Bishop was aware of the possibility that slate could be carried out of Blaenau south and eastwards, taking most of the traffic away from the sea at Porthmadog. In the 1880s the Board of the Festiniog Railway Company, aware of competition, tried to sell their railway to one of these train companies, but no arrangement was made.

Spooner's Private Inspection Cart

On 12 February 1886 Charles Spooner was involved in an accident. He had designed an inspection cart that resembled a small boat. It had a curved prow so that it could open railway gates. This day, he entered his cart in Blaenau along with two ladies who were relatives of his. He had neglected to take up the single-line staff, which would have given him permission to enter the line. A train was coming up the line and near the Moelwyn tunnel, they collided. The boat was smashed to pieces, and one of the ladies, and Spooner himself, was badly injured. A year later, he was replaced

In 1886 Charles Spooner had an accident in his specially built rail carriage, the *Boat*. This was worked by gravity and had a brakeman. Notice the prow, designed to open gates. Spooner was badly injured in the accident and he died in 1889 at his home Bron-y-Garth, Borth-y-Gest.

as secretary to the company by J. S. Hughes. He died at his home, Bron-y-Garth, in 1889, aged seventy-one.

Charles Spooner was buried in Beddgelert and his funeral procession was accompanied by quarrymen and railwaymen. He may have been a prickly man, perhaps overbearing and eminently Victorian, but the community knew the contribution he had made to their economy and society. Unfortunately, his board of directors at the Festiniog railway showed little warmth or appreciation. Contrasting with the fulsome praise he received in the local papers, the board acknowledged his passing with a note in the minutes.

He may have been a hard taskmaster, but he left a legacy of efficiency; the railway ran like a well-oiled machine. There were eight passenger trains every day, each way.

The quarrymen still had to endure fairly Spartan conditions, but the first-class passengers were lapped in almost Byzantine luxury, with mirrors in their compartments, morocco leather seats, carpets, antimacassars changed daily, and copper foot-warmers in the winter. (John Winton, page 71)

The company published a booklet *Rules and regulations* which applied to all work on the railway. It was steeped in routine. At 6 a.m. a bell or whistle signalled the start of work; the working day ended at 5 p.m., except on Saturdays when work stopped at noon. Workmen who were late for work had their wages docked by amounts corresponding to the time lost. The men worked a 55.5-hour week. From 8 a.m., for half an hour, they took breakfast. Workers could work overtime providing they had already worked full hours, with extra pay corresponding to half an hour every 3 hours. Sackings were common. Men could be dismissed for entering the works by any other gate than the timekeeper's. Beer or spirits were entirely banned and anybody seen with those on railway land were sacked immediately. Material belonging to the railway was not to be taken out, on penalty of dismissal. Work was subject to one week's notice, either way.

20

Steaming Ahead

The years 1877 to the end of 1889 (when Spooner died) were very busy and constructive years for the Boston Lodge manufacturing unit.

October 1875 saw drawings by G. P. Spooner created for the new engine No. 9 *Taliesin*. It took the No. 7 when *Little Wonder* was withdrawn.

A minute book dated 14 January 1877 read: 'that a new bogie engine be built and that an erecting shed be made to accommodate the work.' This engine was No. 10 *Merddin Emrys* and it was completed in 1879. Boyd's words are, 'It is believed to have been constructed entirely, including boilers, at Boston Lodge.' (p. 147)

A rare photograph of the Boston Lodge engineering works in the 1890s. William Williams is in the centre, his arm raised.

Double Fairlie locomotive " Merddin Emrys " as first built. The picture shows the original wheel-operated regulators

Double Fairlie locomotive " Livingston Thompson " (now " Taliesin "). The final form of the Festiniog double Fairlie type

THE DOUBLE-BOILER FAIRLIE ARTICULATED LOCOMOTIVE AS DEVELOPED IN THE BOSTON LODGE WORKS OF THE FESTINIOG RAILWAY

Merddin Emrys and *Livingston Thompson*, built in Boston Lodge in the years 1879 and 1885 respectively.

In July 1879 George Percival Spooner (son of Charles Easton Spooner) was made locomotive superintendent at a salary of £100 per year.

At this time, the manufacturing unit was at its full strength. Headed by William Williams and Percy Spooner, the former applying his locomotive running and manufacturing expertise, with the latter doing the desk work as chief designer, it was an ideal managing unit. The manufacturing side included Williams's brother, John Williams, and two of Williams's sons. John Williams's grave in Holy Trinity church, Penrhyndeudraeth, displays the Welsh word Peirianfardd, which means 'poet of the machine'. *Merddin Emrys*, which was larger than the *James Spooner*, came first, then its sister engine the *Livingston Thompson* (first mooted in a minute of August 1882, authorising the building of a locomotive to replace *Little Wonder*). This was named after the company's chairman. This engine is described by Boyd as,

> G. P. Spooner's last masterpiece (though he was abroad when it was completed), the last double Fairlie built for use in Great Britain, and in tribute to the workmanship and maintenance afforded them and since, both these Boston Lodge-built engines survive to the present day, having earned their cost many times over. (p. 133) [Boyd was writing in 1955]

William Williams succeeded as locomotive superintendent in 1881.

Building the *Merddin Emrys* was the biggest job undertaken at Boston Lodge to date. Boyd writes (p. 129),

> Having in mind the facilities of the Works, however suitable for heavy repairs they might have been, the construction of such an engine can only be judged as a remarkable feat. It is known that the boilers were completely fabricated on the spot, and the whole enterprise speaks well of the skill of the men under G. P. Spooner's direction, who designed the engine. Larger than the *James Spooner*, it was the final type of double engine, furnished and fitted with all that experience of running the previous two could give.
>
> As built *Merddin Emrys* had Spooner's patent wheel-operated regulators, which could be worked by handle or wheel, with engines either double or singly connected thereto, wagon-top boilers, stove-pipe chimneys and a longer total wheelbase than before used. The centre portion of the cab was not roofed.

George Percival (Percy) was involved with the close activities of the railway at an early stage. It began about 1875 with the designs for the Fairlies. His father had laid the foundations, but always regarded himself as a civil engineer, having parts to play in the design and growth of narrow gauge railways in different parts of the world. One of his projects was the laying out of the 2-foot-gauge Darjeeling–Himalayan line in India. Charles Spooner had a three-masted schooner, the *C. E. Spooner*, built in Porthmadog, named after him.

One of the problems with the double-Fairlie design originally lay in flexible pipes for carrying steam to and from the cylinders when the engines were swivelling; they were prone to breaking under twisting pressure and so lost power. The original system was replaced by William Williams and his engineer team, who made the connections steam-tight by using spherical steam connections. These engineers also added weights at the ends of the bogies to counterbalance the cylinders. Early Fairlies had a tendency to be unsteady on the rails and were prone to derailment.

Merddinin Emrys and the *Livingston Thompson* carried rounded sandboxes with a lipped lid featuring a prominent central knob, designed by John Wiliams, the blacksmith and ironworker who had a workshop at the Porthmadog quayside. He also designed a new type of windlass which could be handled by one man for the decks of ships. It is said that he failed to patent the design and it was widely copied, including by shipbuilders on the eastern coast of the USA. He was the poet son of Richard Williams, the original blacksmith of Tremadog, who worked for Madocks in the construction of the second embankment, the Cob. He was William Williams's father's brother; therefore his uncle.

In 1883 a severance took place which was of profound importance to the Festiniog railway, particularly its manufacturing unit. Boyd writes,

G. P. Spooner committed an indiscretion with one of their servants and a child was born. His father, a very sedate man, was considerably shaken by this and G. P. Spooner was packed off to India in 1883. He returned later and had lost much of his money; he certainly inherited the engineering abilities of his father but not his business ability. He became a Special Constable at King's Cross during the First World War and died as a result of a stroke on 21st January 1917. After G. P. Spooner went to India, his father relied on William Williams, Works Manager at Boston Lodge. (p. 134)

On 11 October 1886, three years after the Percy incident, William Williams showed his respect for his erstwhile working partner by sending his son John (Johnny) to join Percy in India. He had much work on his hands in designing new railways and supervising their construction, so he needed experienced railway men. On 14 October Percy sailed from London. John Williams became chief executive of the East Bengal Railway; after his retirement he returned to his home area and is buried in the old Anglican church in Penmorfa. His son, Edwin, continued in India under the Raj, working in forestry: he retired and went to live in Penmaenmawr with his wife Myrtle.

Finishing a chapter, John Winton writes,

There were always personal tensions, between Samuel Holland and the railway, between the Oakeley family and the railway. There were regroupings and resignations on the Board. Animosities sometimes continued after death. The last connection with the Spooner family was severed in June 1909 when William Williams the locomotive superintendent (and Scribner's 'most trusted mechanic on the line') finally retired. He had served the railway for sixty years. He was the man who built Spooner's garden railway. Williams had disagreements with Frederick Vaughan, then managing director, and he asked that when he died his coffin should not be taken past Boston Lodge on its way from Portmadoc to Minffordd lest the enmity he had incurred should follow him into the grave. It was a curious incident, showing the astonishing power the railway had over the minds of those who worked on it. (*The Little Wonder*, p. 79)

William Williams, known in poetry writing circles as Gwilym Meirion, retired from the Festiniog railway in June 1906, on a pension of £1 per week. He and his wife Elizabeth celebrated their Golden Anniversary on 11 April 1907. They were required to vacate their lifelong home, No. 3 Boston Lodge, and they moved to live in New Street, Porthmadog, where Williams later died. He is buried at Holy Trinity church, Penrhyndeudraeth.

21

The Railway's Decline and Re-emergence

Slate, which is at the heart of our narrative, was difficult to handle financially: it was a boom or bust story. It was necessarily labour-intensive, requiring hand labour in its extraction and treatment. There is no machine that can look at a pillar of slate and decide how to cut it up into roofing slates; this requires the judgment and expertise of a human being. So, unlike many other products which are prepared for market, it was impossible to entirely mechanise the production of slate. Consequently, huge profits were made when the going was good, but when it was not, tens of thousands of pounds could be lost in a short time.

By 1900 the industry was undercapitalized, using obsolete machinery and operating in small and unprofitable units. It was in poor state, too, with increasing competition from imports of foreign slates and a growing tile-manufacturing industry ... In 1902 there was a strike in the Ffestiniog quarries. At the same time, the shipping trade with Hamburg went into a recession which affected the harbour at Porthmadoc. Several of the smaller quarries closed, together with some of the smaller railways.' (Winton, p. 85)

Ffestiniog suffered a drastically reduced output of slate. On 5 August 1914, the government took over the Festiniog railway and part of the Boston Lodge works was taken over by the Ministry of Munitions.

So, the period through the 1890s and up to the outbreak of the First World War was one of very difficult trading for the Blaenau Ffestiniog slate quarries and mines. The three years' strike at the Lord Penrhyn Bethesda quarries saw the rise of worker power and changed the previous relationship between worker and owner. The huge Oakeley quarry/mine suffered serious rock falls. Slate carriage was in decline and the railway had a reduction in income. Employment was declining and when recruitment came before and during the war, thousands of men left North Wales to join up. Coal, mined mostly in South Wales but also available in the north from the quarries in Rhosllanerchrugog, Ruabon and Wrexham, and Point of Ayer in Flintshire, became scarce due to its use in factories servicing the Army, and the Festiniog railway's locomotives lost power because of inferior fuel. The Festiniog railway was short of men and resources and essential repairs and renovations were not carried out. What was once a bustling harbour, with the Festiniog railway at its heart, the pump of the lifeblood of the town and its economy, ran down and decayed.

The desolation of what once was a busy port is well attested by the following extract from *Brief Glory*, a book by D. W. Morgan who sailed his boat *Dewdrop* just after the war:

But Portmadoc! As the current carried the Dewdrop into what would have been, when I last saw it, a busy and crammed-full port I behold desolation. Not a ship alongside Madocks' erstwhile new quay, no engine puffing along the narrow-gauge lines, no little pointsman in uniform with his dark-shadowed eyes, no ship-chandler or Puddin-rice shop. I steered my boat over water full of shadows until my bowsprit almost touched the bridge, and here I moored, feeling like an intruder in some hallowed graveyard and wanting to speak in whispers.

A few days only did I linger in the uncanny stillness. After visiting a few well-remembered spots I fled, fled as fast as wind would speed me; nor did I ever revisit the old derelict port in any ship, large or small.

The First World War, that major tragedy of history, killed so many, principally a generation of young men who would have gone on to renew and replenish the society of Great Britain, taking forward the genes of their ancestors to create one of the most successful countries in the world. As it was, it was left to the non-serviceman and the war's survivors to take up the challenge of their weakened society. Women also significantly contributed to the war effort.

The Festiniog railway was a *leit motif* of this. It was diminished by the war. It lost much of its *raison d'etre*, the transport of slate, although passenger numbers were kept up. The locals loved their 'lein bach' and they took a trip in it whenever the fancy took them. From Blaenau, they came down to port for their groceries and other domestic goods. Schoolchildren stepped on it lower down and it carried them up to the high-quality Ysgol y Moelwyn secondary school in Blaenau, through a magic land of lakes, cuttings, long and short curves, a dark tunnel and rain-dappled trees and leaves, and all the while accompanied by the unmistakable smell of engine smoke and steam, and the haunting cadence of the locomotive's horn.

In the early 1920s, it was revealed that the Festiniog railway was spending significantly more than it was receiving. A character called Major Holman Frederick Stephens from Tonbridge, Kent, (known in railway circles as Colonel Stephens) took over the management of the line.

The North Wales Narrow Gauge Railways business had been created in 1878 to build a narrow-gauge railway between Porthmadog and Caernarfon: the first section of the proposed line was constructed as far as Dinas, 3 miles south of Caernarfon. This project continued for forty years and ceased in 1916. A section of the line, built with enormous engineering difficulties passed through the Aberglaslyn Pass from Beddgelert, and was virtually finished by 1908, when it became exhausted. After much public demand, in 1923 the line, now the Welsh Highland, had been restored, a track laid across the high street in Porthmadog, and public traffic started. One of its problems was that it extended only to Dinas, but passengers to Caernarfon, 4 miles distant, had to then take a bus. Another problem was that the larger Welsh Highland engines did not easily fit the smaller dimensions of the Festiniog railway. In July 1924, a train of Welsh Highland and Festiniog coaches drawn by Festiniog's *Palmerston* and Welsh Highland's *Moel Tryfan* became separated in the Moelwyn tunnel; for over an hour, passengers sat in the dark while a coupling was repaired.

In July 1934, the Festiniog railway took a forty-two-year lease on the Welsh Highland for a rent of £1 a year. It was a very bad acquisition. In September 1936, the Highland closed for passenger traffic and in 1941 the rails were sold for scrap.

The later history of the Welsh Highland railway is long and complex, involving the concern of Gwynedd County Council, litigation and a search for the existence of shares. However, it came about that the Welsh Highland railway and the Festiniog railway came together to rebuild a line to Caernarfon, and since 1988 the Festiniog railway has been working on it. A first section was opened in 1997 and by 2010 the tracks of the two railways had been reconnected at Harbour Station, with passenger services linking Porthmadog to Caernarfon starting in 2011.

Just as the First World War had done, the Second World War again diminished the Festiniog railway. Again there was a shortage of staff, and slates were not part of the war effort. There was much maintenance to be done and the track route was in neglect. Passenger services were suspended in 1939. On 1 August 1946, the last train ran up the line. Succinct letters of dismissal were sent out to the staff.

As John Winton brilliantly wrote (*The Little Wonder*, p. 110):

The Company were thus caught in a grip of an impasse which would have made old Samuel Holland chuckle. The railway could not be bought or sold, and it could not be leased. Those who owned it did not want it, but could not get rid of it. Those who wanted it could not afford to buy it. It was not a commercial proposition, and never would be. Nobody would take on the task of running it, because it was worth more as scrap than as a railway. But it could not be sold as scrap.

In 1954 Alan Pegler entered the struggle. He was a railways man, a member of the Board of the British Railways eastern region. He bought and took the *Flying Scotsman* to the United States. On 21 September 1954, the first journey since its closing was made on the railway, drawn by a Simplex engine across the Cob. Morris Jones, the foreman fitter who had last worked on the railway in March 1947, rejoined the staff and continued work on restoring the locomotive *Prince*. Robert Evans worked on the railway for sixty years, for almost twenty-five years as manager, and retired in June 1955 when Allan Garroway took over as manager. On 5 March 1955 the first engine and coaches came in to Blaenau Ffestiniog.

In 1954 the Central Electricity Generating Board originated a scheme for flooding part of the northern end of the line. The Festiniog Railway Company sued for compensation and what followed was the second-longest legal battle in British legal history, taking eighteen years and two months. Finally, in 1972, compensation was agreed.

A 2.5-mile deviation was constructed in order to bypass the Ffestiniog hydroelectric power station and its reservoir (Llyn Ystradau). The original Moelwyn tunnel was plugged at its northern end. A new tunnel, 310 yards long, was constructed by three Cornish tin-mining engineers with a team of employees and the work of these 'Deviationists' is still fondly remembered.

The Festiniog railway fully reopened in 1982 and has become the second-largest tourist attraction in Wales after Caernarfon Castle.

The second half of the century saw the Festiniog railway, helped by bands of volunteers, overcome one obstacle after another, until, by completing its route to Blaenau Ffestiniog it became what it is now (along with its sister railway the Welsh Highland railway), well-run, attractive, safe and profitable. The workshops of Boston Lodge built two double Fairlies; the *Earl of Merioneth* in 1979 and the *David Lloyd George* in 1992.

Combined, the Welsh Highland railway and the Festiniog railway is the UK's longest heritage railway at 25 miles from Caernarfon to Porthmadog. Three of the original Festiniog railway locomotives are still in service. Passenger numbers of both railways have increased steadily, as promotion and word of mouth brings in an increasing number of visitors.

One final point worth mentioning is that originally vehicles intent on entering Porthmadog from the south had to pay a toll at Boston Lodge. This was taken into public ownership in 2003 when the Welsh Assembly purchased the rights, so now there is no toll.

Linda preparing for another Up-train trip to Blaenau Ffestiniog at the harbour station, Porthmadog. Photograph taken in 2015.

The *Earl of Merioneth* coming over the Cob about to enter the station in Porthmadog. Photograph taken in 2015.

22

The Manufacture, Launching and Seagoing of the Vessel *Pride of Wales*

The process of manufacturing involves people and workers, and the success of the process frequently depends on their skill and commitment. There is no doubting these qualities in the working people of Porthmadog in the nineteenth century. They worked with pride, keen to make an excellent product, and to benefit their community, its self-assurance and well-being both socially and economically. They worked very hard to make the slate industry successful. They manufactured rolling stock for their railway and built on their own shores efficient and profitable sailing ships. It was all, essentially, hand work; they were artisans, but some of them developed high-level skills. This category included the ship designers who led teams of shipbuilders – one of the most respected being Simon Jones, who built his ships mostly on the eastern shore of Borth-y-Gest.

The year 1870 was an important one. February saw the Festiniog railway come into its own with the trials of its extraordinary double-Fairlie locomotive, the *Little Wonder*. Attended by a representative of the Imperial Russian Commission and other important guests, this little locomotive hauled a long and heavy train up steep gradients and around challenging corners, and was a resounding success. It was the high point of Robert Fairlie's professional life. Also on display was William William's miniature steam engine, *Topsy*, running on tiny rails around the Spooner's garden at Bron-y-Garth, looking over to Borth-y-Gest. And at the same time, Simon Jones was building his masterpiece of a sailing vessel on a site hardly more than 400 yards from Bron-y-Garth. We can hardly have a more serendipitous time and place. The manufacturing of locomotives and sailing ships came together in time and place, representing the best that Porthmadog manufacturing could offer.

We can generalise about industries, give data, identify high and low periods, say where they occurred, and even attempt to invoke what it was like to be involved, but nothing comes close to having an account of the process of manufacture by a person who was actually there. We have no chronicler from the slate and railway industries of Penrhyn and Porthmadog, but we are very fortunate to have one from the world of Porthmadog/Borth-y-Gest maritime history. Henry Hughes was born into the extensive and colourful world of Porthmadog shipping. He saw the harbour grow, slate carried and stacked at the quayside, ships being overhauled and brought in from other ports and countries to utilise the skill of local artisans, ships being built on the muddy bank of Rotten Tare and the whole seaside voyaging experience, for he served in the ships, sailing in them to distant places. His books *Immortal Sails* and *Through Mighty Seas* are beautifully written and are an important and authentic record of local maritime history.

We cannot do better in recounting Porthmadog and Borth-y-Gest's contribution to this history than to go to James Henry Hughes's first-hand account (in *Through Mighty Seas*) of the making and sailing of the barque *Pride of Wales*. (An extract is reproduced by permission of a relative of the author). He was Emrys Hughes's brother and he wrote his two wonderful books in the 1940s. He begins;

The heroine of this book is one of those tiny ships, one of the enormous fleet that sailed the seas little noticed or known. Few of them carried more than five hundred tons. Their length rarely exceeded a hundred and forty feet, and their deck clearance to the sea (free board) could be gauged in inches ...

[p. 14] In 1868 David Morris announced that he was going to build a vessel half as big again as the largest ever built in the place, not only the largest in tonnage but the most ambitious in rig and detail ... he had realized the needs of the future. He had an intuition that there was a place in the world for his dream ship. He visualized golden days by the coral strands of India. He gambled his all on the venture ... The ship was to be called the *Pride of Wales* ... Unruffled and determined, David Morris was found in constant and close consultations with Simon Jones, the designer who, for years, had yearned for an opportunity to put his undoubted artistry into full effect.

The keel was laid at Borth-y-Gest ... [there was] the problem on conveying the huge oak timbers, necessary for the work, along unmade and narrow roads ... But the stout hearts of eight heavy-draught horses turned the difficulty into a local entertainment ... Crowds used to gather at the foot of the steep hill known as Pen-y-Clogwyn for the sheer delight of seeing the great oak trees being galloped up this narrow ravine. The jangling chains of harness, the cracking of whips, the clatter of hoofs, and the encouraging shouts from the horsemen filled the air, while the villagers, hidden by the dust, revelled in the fun and pushed their utmost behind the fallen giants. Spars and sails, tackle and tanks, ropes and rigging, followed in an endless treck. Soon the wonder ship began to take shape. The gaunt hull towered above a host of pleasure craft and fishing boats which speckled the beach. Her elegant stern dipped over the tidal waters of the Atlantic. Her high bows peeped proudly over well-timbered meadowland undulating towards the shorn and rugged Moel-y-Gest which clouded the sky a mile away northwards ...

When the people of Portmadoc did see the figure-head they were very proud of it. A great deal of pains and craft had gone in to it. It represented a young girl of seventeen moving forward as it were with a sprightly gait. A wealth of waving hair covered her head, and it appeared to be ruffled and freshly blown by the winds of the sea. Her aquiline features were wreathed in a seraphic smile. A strand of coral adorned her neck, with pendant ear-rings in keeping – a symbol of the sea. Her dress was white and short, displaying shapely legs which, with their buckled shoes, rested on an elaborately-carved plinth tinged with gold and green. The right arm stretched over the sea, and in the hand was a red rose, an offering, to greet the sun or to humour the storm.

The launching day created a great deal of excitement, and was treated as a local holiday. Borth-y-Gest turned gay. There were flags on every house, and the main street was festooned as though for a coronation. Every vantage point carried the slogan *'Good luck, Pride of Wales'*. Shores and scaffolding were cleared away so that an uninterrupted view of the ship was available to all for the first time. Her lower masts and bowsprit only in position were festooned by gaudy code flags. A silken burgee thirty feet long, emblazoned with her name, flew from a staff on the main. The elaborate carving of sinuous reptiles in intertwined confusion, which completely covered her stern, was then unveiled for the first time.

All the sages were agreed that the ship was a joy to look upon, and that Simon Jones had reached the zenith of his career. Painted a rich bottle-green, girded with a gold strake, the Pride of Wales looked very much like a beautiful toy in a shop window.

Then the climax ...

The much-discussed figure-head was still shrouded in linen, but as soon as a tall young lady mounted the platform to perform the naming ceremony and to send the bottle of wine crashing against her bows, the shroud was torn away, and the figure-head revealed. The news flashed round: 'It is Miss Morris, Miss Jenny Morris, the one who is sending the Pride of Wales gliding smoothly to the sea.'

While the Pride of Wales was loading her first cargo in Porthmadog, it became known that she was to sail at once into the sunny seas of prosperity. A long charter under the Indian Government was signed and sealed, and as she sailed blithely through Portmadoc bay, her proud owner in command, she was waved a long farewell by crowds of fond admirers.

For years the Indian Ocean carried her successfully ... [she] brought credit to her distant home as she bowled along with the regularity of a mailboat between Rangoon and Chittagong ... The commander of a Liverpool full-rigged ship, one Captain Jones, of Bank Place, Porthmadog, told me that in the early seventies they sighted a ship in the Indian Ocean. She was wallowing amid spume and spray under a terrific pressure of canvas and moving rapidly through the sea. Twenty-seven sails, ivory white, and fitting like buckram, scooped up the strong breeze of a tropical day. 'What ship is that?' asked the full-rigger. 'J. Q. P. N.' (which was the code number of the Pride of Wales of Portmadoc) came the reply. 'What's the hurry, David?' was the next question from the Liverpool ship, followed by her recognition numbers. Both captains came from Portmadoc and were old friends, but had not met for years. There was a friendly wave of the hand as the ships passed on different courses ...

Fifteen years afterwards, when I joined the ship, I was able to read the logs of previous years ... After many years in the East Indies the Pride of Wales sailed for England. She had paid for herself many times over in this while. Having had good fortune with his other ships, although they had difficult names to live up to, such as the Success, Ocean Monarch, Royal Charter, and Excelsior, the owner and commander retired at a very early age. The new commander, Evan Pugh, sailed the Pride of Wales hard for the next ten years and visited every part of the world. In 1887 Captain John Griffith took over the command and held it until she foundered. In 1889 she left Columbo for Havre loaded with coffee, and as she was due for a Lloyd's re-classification in 1890 she retired to her native waters of Portmadoc for the necessary refit ...

I raced down to the quay to inspect the Pride of Wales from a closer range. All the marine growth and barnacles of two oceans seemed to have stuck to her sides. Her bottle-green timbers were like Joseph's coat of many colours. The gold strake and the elaborate carvings around her stern were shabbily tarnished. The beautiful figurehead was sadly in need of a wash and brush up. Her decks were white enough, but the bulwarks, boats, and deck fittings, were more rust than paint ... One Evan Lewis, a local carpenter, said she looked the part only a fine ship could have played, and that some of his physic would soon bring her round ...

THE CALL OF THE SEA

The refit started in October, and was vigorously continued through the winter. By February, 1890, her seams had been entirely recaulked. New parts had been grafted on where necessary. Putty and pitch filled the minor lacerations in her sides. They used to say in those days that old wooden sailing vessels were kept afloat by the three P's: Putty, Pitch and Providence. New copper reached to her loading line. New rigging, spars, and sails brought confidence back in her ability to put up a few more rounds against the demon storm.

A new fo'c'sle was built on deck to replace the squalid dungeon low down on the forepeak ... A certain Dr Evans came to my help and persuaded my mother that a few voyages through the tropics would build up my frail frame. Captain Griffith too was invited home to be wheedled to give me a berth in the cabin, and be asked to educate me to the sea.

Thus I went to sea instead of going to school ...

I will now give some details about the ship that was to be my home and mode of transport over many thousands of miles of ocean travel:

THE 'PRIDE OF WALES'
Rig – Three-masted Barque
Length – 125 feet
Beam – 26.5 feet
Draught – 14.5 feet
Tonnage – 298 register
Tons – 500 burden
Deck – Two, full-size poop and main-deck
Cabin – aft
Fo'c'sle and galley – abaft the foremast

The freeboard was three feet. This means that, walking along the main deck amidships, *three feet was the distance to the surface of the sea,* so that it did not do to think too much about that in a storm.

The cabin was cosy. It had mahogany panels with a white ceiling, and a decorative skylight in which hung a tell-tale compass and a swing paraffin lamp and the usual barometer and clock. Leather settees surrounded a mahogany table. There was a snug copper fireplace with a mirror above the mantelpiece. The captain's berth led out on the starboard side. There was an after-cabin which was dark and dingy, lighted only by dead lights. The chief mate's berth led in one direction on the port side, and that of the boatswain and mine in other. The only light that penetrated these bunks struggled through a small deck dead light. I could not see the foot of my bed ...

... In the early nineties, Portmadoc harbour presented a scene of unusual activity. The still atmosphere was broken at an early hour by the metallic clank of the dropping pawls of many windlasses. About a dozen ships were preparing for sea ... The Pride of Wales was the draw ... she had visited the port only once during her life, and was not likely to come again for at least another seven years ... and tailing on to the upper topsail halliard, I pulled my weight to the strain of the rollicking shanty

Santa Anna is going away
Away Santa Ana
Santa Ana is going away
Along the planes of Mexico

The two tugboats *Snowdon* and the *Wave of Life* were busily dashing up and down the harbour, tugging the ships from the shallower muddy quays and sending them on their way. There was little hope of our floating until the top of the tide. When the *Wave of Life* did come to make fast, there was only a short anchor to get in before the order 'Let go the stern rope' was given ... Two dozen hardy sailors manned the windlass, and for a quarter of an hour the air of this quaint and secluded harbour was filled with melody. Rich voices, blending in perfect harmony, echoed round these rugged cliffs ...

With the order given 'Let go the stern rope' we were soon gliding gently away with the dipping of the ensign they saw the last of the Pride of Wales. We were bound for Germany with 500 tons of slates, the largest single cargo ever taken by a Welsh sailing ship from Portmadoc ...

We were making for Falmouth. Our new rigging had got dangerously slack, and there were other minor things that wanted readjusting after a refit ... we passed many stately ships calling at Falmouth for orders. Swinging on their anchors and heaving gently to the incoming swell, marine growth on and below the waistline testified to a hundred days at sea. Peering over their sides in clusters were bronzed and bearded men, meager of mien, of all ages, with, to me, marvellous blue eyes. I was told that I would look like that in a few months ...

Arriving off the Elbe Light Vessel in the hours of darkness, we were rudely reminded that the river had been frozen over since November, for huge ice floes came crashing against our bows, huge and closely packed ...

As soon as it became known that we had been chartered to take a cargo of general merchandise to Rio de Janeiro the entire crew from the Chief Mate downwards refused to sign on. This meant that the Captain and myself alone were left ... someone had told then, and this was true, that there were half a dozen large sailing ships in Rio unable to leave because their crew were stricken with yellow fever and smallpox. This report concerned the summer season, Brazil's unhealthiest period ...

Strange foreign faces were soon on board – Norwegians, Swedes, and Germans, and a very good lot of men they turned out to be. All excellent sailors, each one had an accordion. The late eighties and early nineties were the peak period of the transition from sail to steam, and it was becoming increasingly difficult to man the cumbersome large sailing ships with the right type of seaman ... no reasonable married man could be expected to refuse £4 10s a month in steamships as against £3 in sailing ships. It is difficult to know how these men kept wives and families on that amount. Sometimes the wife was allowed to draw half-pay every month, and she would be very lucky if her husband brought any appreciable amount of the other half back at the end of a voyage.

Little ships like the *Pride of Wales* were preferred by seamen for the simple reason that the work aloft was much lighter. I used to watch gangs of seamen walking about Bute Docks, Cardiff, choosing their ship. They would slink by a four-masted full-rigged ship, and doing so would cast their eyes aloft to see a veritable forest of spars – a Chinese puzzle in ropes. They would then turn up their coat collars, rub their hands, and double for the first barque. I was frequently accosted by them in this manner: 'What are you bound for in this packet?' 'The West Indies.' 'Ah, that's the spot for me.'

The old salt, too, would become very artful. If a ship was bound for Valpariso or 'Frisco, he would work out where he would be in the winter season. He would avoid rounding the Horn at the bad time of year, if instead he could get a ship going in another direction, say to the East Indies.

In 1895, it was no uncommon sight to see the remnants of the great sailing ships run by apprentices and youths. The nation had by then lost one of its finest types of manhood ...

[the run to Rio] ... all I saw during the five thousand miles' journey were the terrifying breakers, angrily tossing on the half-covered Goodwins, with stumps of masts sticking up from a recent wreck, a distant view of the twinkling lights of Folkestone, and later the Eddystone Lighthouse.

The author, Henry Hughes, goes on to tell various tales of his life at sea, including an interesting account of how the island of Aruba in the Dutch Antilles, off the coast of Venezuela, became, through the nineteenth century, such a profitable (and dangerous) area for Porthmadog ships. The ships sometimes went there in a single trip in ballast and came back laden with the valuable phosphate rock, which was much needed in Britain as a constituent in crop fertilizer.

Conclusion

This extract from the writings of Henry Hughes carries information, and an atmosphere, which marked the end of an era, not only at Porthmadog and its environs, but across the UK. The nineteenth century had seen the expansion of the industries of Porthmadog, driven by the value of slate. Blaenau Ffestiniog was the heartland. It contained a material which made fortunes for some and bankrupted others. Extraordinary skills were learned and used on the shores of Tremadog Bay in locomotive and ship manufacture. From 1795, what started as a bleak, gorse-strewn, almost unoccupied, stretch of coast had been converted in the following decades into a place where some of the finest sailing ships in the world were being built, and where locomotive building had been developed to such a level of expertise that here, for the first time ever, steam engines were run on a 13-mile-long commercially viable narrow gauge track over mountain terrain.

Slate, sail and steam have had a symbiotic relationship in the Porthmadog/Ffestiniog axis. They have passed through the conduit of history and have reached the final stage in the development of the Festiniog railway, with, now, the Welsh Highland railway.

Porthmadog's shipbuilding has long gone. Blaenau Ffestiniog's large-scale quarrying of slate has shrunk to a small quantity. And the railway? It has been reborn. Seeing it now, with its new trip to Caernarfon, with its well-run systems, sound management, bright, polished trains, some of them carrying the original names, with passengers enjoying spectacular scenery, is a happy experience. This goes some way towards rectifying the unhappiness of the past, in the dark side of industrialism, when quarrymen suffered terrible working conditions, when families struggled for a living, and sailors came back from voyages feeling lucky to be alive.

Bibliography

Porthmadog Ships by Emrys Hughes and Aled Eames (Gwynedd Archives, 1975).

Madocks and the Wonder of Wales by Elizabeth Beazley (Faber and Faber, 1967).

Portmadoc and its Resources, 1856 by Madog ap Owain Gwynedd (Delfryn Publications, 2013).

Narrow-Gauge Railways in North Wales by Charles E. Lee (The Railway Publishing Co, 1945).

The Festiniog Railway, Vol 1 1800–1889 by J. I. C. Boyd (Oakwood Press, 1965).

Portrait of North Wales by Michael Senior (Robert Hale, 1973).

Ships and Seamen of Gwynedd by Aled Eames (Gwynedd Archives, 1976).

Sails on the Dwyryd by M. J. T. Lewis (Snowdonia National Park Study Centre, 1989).

Caernarfonshire Sail by Owen F. G. Kilgour (Gwasg Careg Gwalch, 2008).

Through Mighty Seas by Henry Hughes (T. Stephenson, 1975).

Immortal Sails by Henry Hughes (T. Stephenson, 1969).

The Gestiana by Alltud Eifion (Delfryn Publications, 2013).

The Little Wonder by John Winton (Festiniog Railway Company and Michael Joseph, 1975).

Slate Quarrying in Wales by Alun John Richards (Gwasg Carreg Gwalch, 2006).

Gazeteer of Slate Quarrying in Wales by Alun John Richards (Gwasg Carreg Gwalch, 2007).

How Festiniog Got its Railway by M. J. T. Lewis (Railway & Canal Historical Society, 1968).

The Memoirs of Samuel Holland (Merioneth Historical Society, 1952).